Divergent Tracks

Divergent Tracks

*How Three Film Communities
Revolutionized Digital Film Sound*

Vanessa Theme Ament

BLOOMSBURY ACADEMIC
NEW YORK • LONDON • OXFORD • NEW DELHI • SYDNEY

BLOOMSBURY ACADEMIC
Bloomsbury Publishing Inc
1385 Broadway, New York, NY 10018, USA
50 Bedford Square, London, WC1B 3DP, UK
29 Earlsfort Terrace, Dublin 2, Ireland

BLOOMSBURY, BLOOMSBURY ACADEMIC and the Diana logo are trademarks of
Bloomsbury Publishing Plc

First published in the United States of America 2021
This paperback edition published 2022

Copyright © Vanessa Theme Ament, 2021

For legal purposes the Acknowledgments on p. xii constitute an extension
of this copyright page.

Cover design: Namkwan Cho
Cover image © Miramax Films/Photofest

All rights reserved. No part of this publication may be reproduced or
transmitted in any form or by any means, electronic or mechanical, including
photocopying, recording, or any information storage or retrieval system, without prior
permission in writing from the publishers.

Bloomsbury Publishing Inc does not have any control over, or responsibility for,
any third-party websites referred to or in this book. All internet addresses given in this
book were correct at the time of going to press. The author and publisher regret any
inconvenience caused if addresses have changed or sites have ceased to exist, but
can accept no responsibility for any such changes.

Library of Congress Cataloging-in-Publication Data
Names: Ament, Vanessa Theme, author.
Title: Divergent tracks / Vanessa Ament.
Description: New York : Bloomsbury Academic, 2021. |
Includes bibliographical references and index.
Identifiers: LCCN 2021000586 (print) | LCCN 2021000587 (ebook) |
ISBN 9781501359224 (hardback) | ISBN 9781501359217 (ebook) |
ISBN 9781501359200 (pdf)
Subjects: LCSH: Sound motion pictures–History. | Sound–Recording and
reproducing–History. | Motion pictures–Sound effects–History.
Classification: LCC PN1995.7 .A46 2021 (print) | LCC PN1995.7 (ebook) |
DDC 791.4302/4–dc23
LC record available at https://lccn.loc.gov/2021000586
LC ebook record available at https://lccn.loc.gov/2021000587

ISBN: HB: 978-1-5013-5922-4
PB: 978-1-5013-7853-9
ePDF: 978-1-5013-5920-0
eBook: 978-1-5013-5921-7

Typeset by Newgen KnowledgeWorks Pvt. Ltd., Chennai, India

To find out more about our authors and books visit www.bloomsbury.com
and sign up for our newsletters.

Contents

Plates	vi
Preface	viii
Foreword by Richard L. Edwards	x
Acknowledgments	xii
1 Introduction: Sound Design as Creative and Cultural Artifact	1
2 Geographical Cultures and Technological Tendencies	25
3 "Viscous Was the Word of the Day": The Interiority of *Barton Fink*	49
4 "How Would You Like to Work on a Monster Movie?": The Expressionism of *Bram Stoker's Dracula*	71
5 "The Sound of the Desert Is Tape Hiss": A Study in Contrasts in *The English Patient*	99
6 Conclusion: Reassessing Sound Design as a Collective Endeavor	121
Appendix A Workflow Diagram for *Barton Fink*	131
Appendix B Workflow Diagram for *Bram Stoker's Dracula*	133
Appendix C Workflow Diagram for *The English Patient*	135
Appendix D Sound Department Credits	137
Bibliography	143
Index	153

Plates

1. Supervising sound editor of *Barton Fink*, Skip Lievsay
2. Barton Fink (John Turturro) as he rings the never-ending hotel bell in *Barton Fink*
3. Barton Fink (John Turturro) enters the elevator, which incorporates Foley to emphasize its decrepit nature in *Barton Fink*
4. The tight suction sound of Barton's (John Turturro) hotel room door adds to the sense of seclusion in *Barton Fink*
5. Barton Fink (John Turturro) attempts to reattach the dripping wallpaper in *Barton Fink*
6. Charlie (John Goodman) and Barton (John Turturro) watch the wallpaper in Barton's room peel away from the wall in *Barton Fink*
7. Studio head Jack Lipnick (Michael Lerner) bangs his desk, which results in the signature "hubcap" sound effect in *Barton Fink*
8. Supervising sound editor of *Bram Stoker's Dracula*, David E. Stone
9. The carriage driver's arm reaches out for Jonathan Harker (Keanu Reeves) in *Bram Stoker's Dracula*
10. The Count's castle gate, which evokes sounds of animals and creatures, as it clangs shut in *Bram Stoker's Dracula*
11. The Count's shadow travels around the room to greet Jonathan Harker (Keanu Reeves) in *Bram Stoker's Dracula*
12. The Count (Gary Oldman) picks up Mina's photo, casting a shadow of ink onto it in *Bram Stoker's Dracula*
13. The Count's shadow jealously attempts to strangle Jonathan Harker (Keanu Reeves) in *Bram Stoker's Dracula*

Plates vii

14 Dracula, as the Wolfen (Gary Oldman), sees Mina (Winona Ryder) as he rapes Lucy (Sadie Frost) in *Bram Stoker's Dracula*
15 As the Wolfen, the Count (Gary Oldman) appears as his young self and clairvoyantly communicates to Mina "Do not see me," in *Bram Stoker's Dracula*
16 The Young Count (Gary Oldman) clairvoyantly communicates to Mina (Winona Ryder), "See me now," at the Cinematograph in *Bram Stoker's Dracula*
17 Mina (Winona Ryder) sees the young Count (Gary Oldman) then continues on to the Cinematograph in *Bram Stoker's Dracula*
18 Supervising sound editor of *The English Patient*, Pat Jackson
19 Rerecording mixer of *The English Patient*, Mark Berger
20 Hana (Juliette Binoche) begins to play Bach on the piano found covered in the monastery, in *The English Patient*
21 Caravaggio (Willem Dafoe) taps the vial of morphine as Hana (Juliette Binoche) plays piano in *The English Patient*
22 Kip (Naveen Andrews) discovers then defuses the bomb hidden in the piano: "Move that, and no more Bach," in *The English Patient*
23 Katharine Clifton (Kristin Scott Thomas) counts the stars in the sky while Almásy (Ralph Fiennes) points to the sandstorm approaching, in *The English Patient*
24 Members of the caravan rush to their jeeps during the sandstorm in *The English Patient*
25 Katharine (Kristin Scott Thomas) and Almásy (Ralph Fiennes) discover romantic tenderness inside their jeep during the violence of the sandstorm in *The English Patient*
26 Kip (Naveen Andrews) defuses a bomb while soldiers in military tanks approach in *The English Patient*
27 Sergeant Hardy (Kevin Whately) tries to stop the tanks from crossing the bridge while Kip defuses the bomb in *The English Patient*

Preface

This book tells the story of how three creative artists—a bass player, a documentarian, and a young animator—who had become innovative sound editors for film, collaborated with their sound film professionals to design award-winning soundtracks during a critical decade in film sound—the 1990s. As movie sound was beginning the awkward transition from editing on magnetic film to utilizing nascent computerized systems, Skip Lievsay in New York, Pat Jackson in the San Francisco Bay Area, and David Stone in Hollywood supervised three critical films that illustrate the technological disruption as a creative challenge particular to the aesthetics, workflows, and political economies of their specific film communities.

Lievsay, a musician turned New York sound editor, found his artistic choices somewhat limited by analog editing practices, and chose to leap to digital editing to allow more creative control over his sound design. Jackson, a public television and documentary professional in San Francisco, began working with Walter Murch, and through the more fluid filmmaking practices of the Bay Area navigated multiple digital sound technologies to create masterful sound design. Stone, an animator who transitioned to sound editing, stumbled into a labyrinth of various digital technologies that did not interface but were massaged into a cohesive and expansively artful sound design.

These three sound professionals, from three culturally unique film communities, along with their colleagues, collaborated on award-winning sound for their films while adapting the technologies to their specific needs.

My method for such an undertaking requires a deep dive into industrial ethnography in film sound. As I was in postproduction sound for several decades prior to writing my first book, *The Foley Grail* (2009), I have a particular expertise in the culture of the film sound professional. It is with this past experience—as one who was in and of the culture—that I bring to sound studies a different perspective, and one that I hope assists in disabusing, once and for all, the notion that sound professionals are technological workers. It is my purpose, more than giving reportage about the technological transition, to shed light on exactly what makes these professionals creative artists. Sound editors, mixers, and Foley artists, in addition to all who contribute to the sound design of a film, are creatives, and not button pushers or machine operators. They use and adapt the tools to create.

The transition of sound editing in film from analog to digital is not a story of technological determinism. Rather, it is the story of determined artists who adapt the tools they are given for their own purposes. Their purpose, as these three stories will persuade, is to create artful storytelling through sound, be the film an offbeat New York comedy, a mythological horror film, or an historical love story. The three films, *Barton Fink* (1991, Lievsay), *Bram Stoker's Dracula* (1992, Stone), and *The English Patient* (1996, Jackson), illustrate three different artistic sound aesthetics, from three different film communities, demonstrating three different approaches to adapting the awkward transitions from analog to digital sound editing.

Foreword

> New technologies radically changed production, distribution and exhibition [in the 1990s]. *Forrest Gump* (1994), one of the biggest hits of the decade, ably exploited computer-generated imagery, so much so it made possible, even inevitable, a future cinema in which location shooting and live production might become obsolete. More and more multiplexes were equipped with one or another variant of digital, Dolby sound. The result was not only higher-quality film exhibition but an increasing production emphasis on action-based sound effects (*whammies*, producer Joel Silver calls them), pop music packages, and MTV video and recording company tie-ins.
>
> (Jon Lewis, *The End of Cinema As We Know It*)

As Jon Lewis highlights in the above quote, all aspects of American filmmaking in the 1990s experienced a radical change due to the emergence of newly available digital technologies. As it would affect other creative industries and their modes of production, fin de siècle Hollywood had to confront the changes being wrought by the digital revolution. The digital transformations that impacted the visual design and editing strategies of Hollywood filmmaking have been well explored elsewhere; however, the digital impact on postproduction sound during this period of transition has been less analyzed and less well understood. Moreover, in the 1990s, the time was ripe for innovation in film sound as theatre owners upgraded their multiplexes for dynamic and multichanneled exhibition of sound including Dolby, THX, and 5.1 surround sound systems. In tandem with these new sonic capacities for filmgoers, a digital revolution was also happening below the line and behind the scenes as sound designers and their crews adapted to new tools and technologies as part of their creative process.

All whammies aside, how exactly did sound professionals adapt as they moved away from long-standing and proven analog approaches to the uncharted possibilities of digital technologies as part of their sound recording, remixing, and editing practices? What kinds of challenges and opportunities did early digital technologies present to these creative individuals working on film sound? What did early innovators in this period of transition discover that paved the way for new approaches to digital cinema in the 1990s? As more and more films utilized digital technologies in the 1990s, the resultant films made a *double* request of the audience: look more closely *and* listen more acutely. The magic of cinema was being rekindled through computer-generated artistry and digitally enhanced films infused our eyes and ears with new moments of wonder and awe.

In *Divergent Tracks*, Dr. Vanessa Theme Ament brings us into the professional milieu of postproduction sound. Through three iconic films, Dr. Ament traces the different ways that sound designers and their crews encountered the advent of digital sound technologies. Perhaps surprisingly, the digital revolution was neither immediate nor handled the same way in different production contexts. This book is a cultural historiography and an industrial ethnography into a period of great change in the way film sound was produced. Dr. Ament carefully traces these changes through three different film communities—those based in the Los Angeles region, the San Francisco Bay Area, and the Greater New York area—and she demonstrates that prior to a new playbook coalescing around digital convergence, there were unexpected early periods of digital divergence and experimentation. Ultimately, *Divergent Tracks* is as much about innovation and the creative process as it is about film and sound design. It is about overcoming technical challenges and making creative leaps; it showcases technological change and highlights aesthetic inspiration; it presents a roadmap to the future of digital cinema and shares a collective drive for inventive and meaningful solutions. Moreover, this book is a reminder that cinema is a total artform and you haven't heard the full story about 1990s American film and the rise of digital cinema until now.

Richard L. Edwards
January 2020

Acknowledgments

This book could not have been possible without the many interviews with the sound professionals who graciously agreed to participate in this endeavor. Over a period of eight years, I have engaged in wonderful conversations with those who are directly credited in this book, and many others who have helped form my awareness to the art and culture of the film industry in the Hollywood/Los Angeles area, the San Francisco Bay area, and the New York/New Jersey area. I am profoundly humbled by the brilliance, talent, and skill I have witnessed over my entire adult life in film. As I have transitioned to the more academic side of film studies, I appreciate the many scholars who have helped me become a better observer of ethnographic and cultural studies, as well as alerting me to the dangers of writing about an area in which I have such familiarity and connection—postproduction sound.

Straightaway, I must thank Dr. Ted Friedman and Dr. Jack Boozer, who led me through a tricky PhD trajectory. Much of this book derives from my PhD studies, and I am certain I would never have understood my place as a writer without these two splendid professors at Georgia State University. I also am grateful to Dr. Alisa Perren, of the University of Texas, Austin, who opened the magical kingdom of media industries and production studies to me. It is because of Dr. Perren that I found my voice in industrial ethnography in general, and sound studies in particular. I would be remiss not to include the inestimable Dr. Rick Altman, who has supported my mission to write about professionals in sound from 2007 onward. Additionally, Altman's prize PhD candidate and now respected sound scholar Dr. Jay Beck is a reminder to me to keep the bar as high as I can, within my own limitations, as he is precise, meticulous, and mindful in all things academic. More recently, I have been fortunate to collaborate with Dr. Richard Edwards, whose depth of film knowledge and devotion to excellence in film scholarship have been remarkable and an inspiration to me. His critiques and notes of my work as I refined the manuscript were immeasurably valuable.

The interviewees I include in this book illuminate the creativity and intellectual thought involved with sound design with their own words better than I could have hoped. For this reason, I must give my appreciation to Mark Berger, Elisha Birnbaum, Lee Dichter, Malcolm Fife, Pat Jackson, David Cohn, Marko Costanzo, Thomas Fleischman, Jim Fulmis, Eugene Gearty, Ann Kroeber, Steve Lee, Skip Lievsay, Tom McCarthy Jr., Bruce Pross, Steve Shurtz, and David Stone, all who added tremendous depth and insight to this book. Their artistry becomes apparent as they speak with passion about their work and describe process, aesthetics, and creative approach, even when describing a technological issue to overcome.

I am eternally grateful to those who pioneered innovations and developed the means by which sound professionals bring their art to our films. This book is dedicated to all the professionals who have worked or now work in film sound, for they bring narrative to life.

Much appreciation to Bloomsbury for enthusiastically shepherding this book through to publication and especially to Katie Gallof and Erin Duffy for supporting the goal of shining a light on postproduction sound and its entry into digital editing.

1

Introduction: Sound Design as Creative and Cultural Artifact

In 1974, when *The Conversation* was released, filmgoers were treated to a new and more conceptually designed use of motion picture sound as Walter Murch introduced an inventive use of environmental and subjective soundscape. Previously, most viewers perceived sound effects as a perfunctory addition to the narrative. This misconception was forever altered by Murch, who created cinematic aurality that began to transform the role of the sound effects editor from that of a craftsperson to that of an artist. Film sound fans and professionals alike began to listen to films differently because Murch—a respected member of the San Francisco Bay Area film community—utilized the available technology to "push the envelope" with sound aesthetics in narrative storytelling.[1] As a graduate of the University of Southern California, and part of the triumvirate of Francis Ford Coppola, George Lucas, and himself, his fresh approach to sound seemed individual and unique. Without the fetters of the studio system to limit him, Murch was empowered to break rules and run free. Working within the era of the "New Hollywood" and commiserating with two film school graduate colleagues who shunned Hollywood trappings, Murch and his colleagues would proceed to reinvent many wheels. This additional explanation is not to minimize Murch's genius, for he surely is gifted with vision and boldness. However, it is to point out that political culture, education, and generational perspective have a place in the breakthrough Murch was positioned to lead. Three men of educational privilege, with confidence and talent, decided to try something new.

Murch, Francis Ford Coppola, and George Lucas were graduate film school students in Southern California. In the 1960s, Coppola attended UCLA, while Murch and Lucas were at rival school USC. The three became close friends and established their own filmmaking culture without the conventions, time constraints, and commercialism of the Los Angeles studio system. Both Coppola and Murch hailed from New York's cultural elite and shared a background in New York theatre and intellectual discourse. Lucas, in sharp contrast, grew up in California's Central Valley, the son of a stationery storeowner.[2]

Murch's freedom to approach sound as an aspect of design in *The Conversation* reflects his relationship with Coppola, his elite education, and his introspective intellectualism. The film as conceptualized by Coppola required more aural "interiority" both for the subject matter and the main character, Harry Caul, than a less thoughtful director might have imagined. Coppola, Lucas, and Murch shared a professional ideal that mirrored their approach to filmmaking in film school as they shepherded varying responsibilities on multiple films. In *The Conversation*, Murch was the film's supervising sound editor and rerecording mixer.[3] This was atypical for a major-release film in 1974. Murch might be yet another well-respected but dramatically lesser-known talent in the Hollywood community, were it not for these three men rejecting conventions during the "New Hollywood" days.[4]

Few would dispute that *The Conversation* was a vital turning point in the history of film sound. Scholars and critics alike have regarded past historical changes in sound design, be they artistic or technological, as significant in the history of film production. *The Conversation* is no less critical, as it set the bar for creative sound work to a new level, which sound designers to this day remember and valorize. This group of young men with film school pedigrees revolutionized the approach to filmmaking in reaction to the preceding filmmaking generation of journeymen and studio-trained craftsmen.[5]

The term "sound design" gives the impression of a planned design, much like production design, set design, or costume design. While this is probably more often true in contemporary times, it has not been the case in the past. The term has been one of dispute among postproduction film professionals from the beginning of its origination. The honorific of "sound designer" came into vogue with Murch, who has used the term to describe his role on various films, such as *The Conversation* (1974) and *Apocalypse Now* (1979).[6] Francis Ford Coppola recalled the term being credited to Murch for his work on *The Rain People* (1969) as his official title, since Murch was not yet in any union, and therefore was not allowed the credit of sound editor.[7] Coppola, however, remembers incorrectly: Murch is credited on that film for sound montage, as he is for *American Graffiti* (1973) as well.[8] The first official credit for sound design is for *Apocalypse Now* (along with sound montage), yet it seems clear that both Coppola and Murch considered the gifted sound maven as deserving of that mantel long before 1979.

Murch revolutionized the art of sound in film at a time when film school graduates were beginning to infiltrate the Hollywood sensibility. Working with Coppola, Murch was able to expand the role of the sound editor to include aural design. However, this concept of sound as a design factor in the overall soundscape of the average film was not quickly adopted. In Hollywood, sound effects were still utilized per the old codified studio system; the more typical use of sound effects tended to be the more prosaic "see a door, hear a door."[9] Although many recall that composer Bernard Herrmann supervised the use of electronic music as a device for sound design[10] in *The Birds* (1963), he was credited as sound consultant, and not as a designer.[11]

Another use of the term refers to a specialist who designs a specific set of sounds for unique use in a film. John Pospisil was the sound designer on *Robocop* (1987).[12] He used various synthesizers to make exclusive sounds that did not previously exist, and is in contemporary terms, a synthesist.[13] By the late 1980s and early 1990s, the infrequent use of the term "sound designer" became a codified title for a few elite sound supervisors. The reasoning behind this expansion was the very real misconception that sound was a craft of knobs and buttons, thus some sound professionals decided to dispel this notion as one discounting the artistry and aesthetics required in the execution of the soundscape. While the film score was recognized as art, so too, the rationale went, should the sound design. To this end, some films began to include "Sound Design" as an official single-card credit at the beginning of the opening film credits, or in the end crawl along with the additional sound crew credits.[14]

Shortly after the expanded sales of "prosumer"[15] equipment, and the correlation of burgeoning film schools, courses teaching "sound design" became the vogue.[16] Presently, the term "sound design" is more ubiquitous than originally intended by Murch, Coppola, or most other film sound professionals. In most films, the sound designer might be the supervising sound editor, or the person who conceives of the overall soundscape. For nascent filmmakers or film students, the sound designer is the person who manages the postproduction sound for the film—whatever might be entailed—regardless if any actual design is required at all. The term is so misapplied that it is often conflated with the film score, which presents another issue altogether: that of the term "soundtrack" as it is commonly used. For the purpose of this book, and in service to all postproduction sound professionals, the term soundtrack will refer to both the film score *and* the sound effects/sound design in concert together. Ultimately, whether constructed to work together intentionally—which indeed is the ideal—or not, they are perceived together in exhibition.

The Shadows Cast by Celebrity Sound Designers

If most film sound fans decide to follow the careers of "sound designers" it would be easy to name some of the few men who are notable that they would list as role models: Walter Murch, Ben Burtt, Randy Thom, and Gary Rydstrom might come to mind instantly. These are the names of sound designers one hears repeatedly. When asked to request an article, engage a keynote speaker, or film an interview, these names are the first that come to mind. If one looks further, the names Mark Mangini, Steve Flick, Skip Lievsay, Ren Klyce, and Richard King might pass the speaker's lips. Why? Because for the most part, those first mentioned are connected to George Lucas—a loyal and supportive producer as well as a master of promotion—and who have displayed tremendous talent, skill, and have won many awards. The next mentioned sound professionals are also highly visible, award-winning, and normally work with high-profile directors repeatedly. Lievsay in particular with the Coen brothers, Klyce with David Fincher, and King with Christopher Nolan will come to mind. These names—all deserving and gifted men—have been minor celebrities and are becoming more familiar to the regular filmgoer. Recently, Mildred Iatrou and Ai-Ling Lee have received attention as the first female sound design team[17] to be nominated twice for the Academy Award for sound design, thanks to the loyalty of director Damien Chazelle. The novelty of their gender in this male-dominated field and Chazelle's meteoric rise to fame have allowed them to become notable more quickly. The question becomes what of the other sound professionals? Who are they and why are they relegated to the shadows of forgotten names who are responsible for amazing soundtracks, but do not have the fame or notoriety? Our celebrity culture and a lack of understanding of the profession itself have led us down the predictable path of valorizing particular individuals.[18] These individuals may not necessarily do something specifically different than other gifted and hardworking sound professionals. However, they have been connected to directors or producers who are in the limelight, so they are, by connection, well known. The problem with this phenomenon is twofold. First, the assumption is that they are the artists and *real* sound designers and the rest of the sound professionals are technical workers. Second, the theories and methods of these professionals are illuminated while the rest of the professionals are somewhere in the shadows, working with less famous directors, or teaching in film schools. Some of these professionals are award winners themselves.[19]

Another mistaken assumption is that sound design is within the territory of the director—that somehow the director has the aural imagination to impart imagery and expertise that is executed by technical workers. Again, this perspective is, with rare exceptions, inaccurate.[20] Directors can and often do collaborate with sound professionals, but it is these professionals who collectively author the soundtrack. While the work may be for hire,[21] directors are unlikely to take credit for sound design unless they are also sound professionals as well. The talents, skills, and training are vastly different. Additionally, whether the filmmaking conventions are from the more Fordist traditions,[22] as they are in Hollywood, the documentary or avant-garde scene, as they are in San Francisco, or the independent film director, which is the predominant model in New York, filmmaking is a collaborative art, and specific expertise and skill sets are necessary requirements for each of the aspects of the art.

The Myth of Postproduction Sound as a Below-the-Line "Craft"

The perception of a below-the-line film professional is one that is best identified with scholarship by John T. Caldwell, who most recently had refined his definition as "all of the workers involved in the 'physical production' of unionized feature film and television who work at fixed hourly rates."[23] Caldwell considers above-the-line to include "'talent' and management."[24] Caldwell's categories derive from the economic and decision-making practices long codified from the historical Hollywood industrial conventions. Film sound professionals—included in the below-the-line category—are often misperceived to be primarily computer-literate technicians. Prior to this misnomer, they were equally misperceived as mechanically literate craftsmen.[25] Neither is actually true. Postproduction sound professionals, in general, are creative artists who use the tools they are assigned to create soundtracks for film and television. The frustration most film sound professionals experience with this false impression is best described humorously by rerecording mixer Kevin O'Connell, who at an event hosted by the Academy of Motion Picture Arts and Sciences honoring the nominees for the coveted sound awards on March 8, 2008, proclaimed when describing his colleagues, "These guys are artists. They call an engineer to replace a light bulb."[26] While O'Connell was joking, he was making a point. The mistaken assumption that film sound professionals—mixers, editors, and Foley artists—are only technical crafts persons and not *also* creative artists continues, as evidenced by their status as below-the-line workers, and by the wealth of research focused on the technology of film sound. The discussion of the designer is often relegated to the reexamination of those few "great men"[27] whose names fall easily off the tongues of film sound scholars and fans. My hope is to shed light on the mislabeling of sound professionals as *only* technical professionals rather than *primarily* creative professionals.

Still today, those who control the environments in which these professionals create the "aural narratives" of films have little—if any—knowledge or regard for the talent and expertise involved in the work these professionals pursue. I use this term "aural narrative" to describe what postproduction sound professionals actually do: create the aurality of the narrative. This disconnect—sound professionals most often being regarded as technical craftspersons and their admitted self-reflexive identity as creative artists—necessitates the question, "How can this misperception be corrected?" As Rick Altman reminds us, silent films were accompanied by music and sound effects.[28] However, once film incorporated synchronized sound, and the process for adding sound was relegated to postproduction, it became an invisible art—so invisible that only those who are enamored with the process even bother to explore how the aural that accompanies the visual is constructed, and by whom. Ironically, the canonical film *King Kong* (1933) and its soundtrack is a fascinating study of the integration of sound effects and film score, included sound effects designed not by a technological wizard but by drummer Murray Spivack, who was charged with managing the music department of RKO studios.[29] Yet, Spivack did what musicians since him have done: figure out how to do what he was asked to do with the technology that was available at the time. As the supervisor of the music department, Spivack was responsible for both the sound effects and the music at RKO.[30] Yet, he is remembered more as a "sound engineer" than as a music supervisor.[31] To put a finer point on Spivack's status as a technician rather than artist, one can look to Peter Franklin's excellent analysis of the sound and music in *King Kong*. Franklin discusses at length Max Steiner's score and its contribution to the narrative. Yet Spivack's role is one of supporting player to Steiner.[32]

David Hesmondhalgh's terminology of "symbol creator" or, as he later refines it, "symbol maker"—one who makes up, interprets, or reworks texts[33]—is a more appropriate description for the postproduction sound professionals and the soundtracks they design. His amplification, "the personnel responsible for the creative input in texts, such as writers, directors, producers, performers . . . ,"[34] works equally well for those who design the aural texts, such as film composers and sound designers. Indeed, Hesmondhalgh expands his exploration to include workers in the music recording industry.[35] Those who design sound for films are creative artists who work in the back end of film by necessity. They have more in common with the film composers and musicians—who *are* considered creative artists—than, for example, the engineers who tweak the insides of the computers and mixing stages where the soundtracks are created. These sound professionals do not merely attach sounds to objects. They create sonic environments, evoke emotions, add synchresis,[36] and alter perceptions with an astute awareness of aesthetics and storytelling. Many sound professionals have a background in music, art, theatre, screenwriting, or some related art form. Few come to film sound from specifically technological training.

These postproduction sound professionals are *primarily creatives* who are also technological workers. To simply relegate this group of experts to a category of below-the-line craftspersons in order to fit a preconceived hierarchy denies the history of film sound professionals and more importantly it marginalizes the work they actually do in film presently. Rather than to concede the delineation of postproduction sound as a craft that is technical and artless, and celebrate the few who prove the rule by being the exceptions, what is more accurate is that postproduction sound is an art *and* a craft—an art that requires original creation and a craft that requires skill and technique—and decries the simple taxonomy that allows for easier accounting and administration. Further, these postproduction sound professionals are artists who utilize and effect changes in the tools they employ, just as artists in other fields do. Rather than be at the effect of the machines and technology, these professionals condition the tools to perform the tasks they require.

Mapping a Digital Transition through Three Film Communities

To expand upon this key assertion, it is useful to examine an important transition in our recent film sound history: the beginnings of the evolution from analog postproduction sound editing to digital. This transition was awkward and took upward of fifteen years. The essence of sound design occurs within the sound editing process and while rerecording is a crucial element in the overall soundscape, not only did the transition to digital mixing happen years later but the sound designer label is one that denotes creation and editing while rerecording denotes mixing, finessing, and blending. Modern technologies have made possible the fusing of these two separate and distinct skill sets, but this was not true during the transition from analog to digital.[37]

My focus is on the era of the 1990s, when the transition was at its most intense. During this decade, postproduction sound began the transition from analog film sound to digital in major motion pictures. While the television industry had begun the adaptation to a digital workflow in the late 1970s and early 1980s, it was not until the 1990s that motion pictures, with all of its complicated moving parts, stepped into the digital transition with any true commitment. The three main American film production communities—Hollywood,[38] New York,[39] and the San Francisco Bay Area[40]—made the initial conversion from analog to digital in the 1990s, but unevenly, and with varying results.

In service to the main purpose, to illustrate the role of sound professional as artist, it is essential to illustrate that the three film communities utilized different digital systems for reasons that were specific to the cultural tendencies of each city. Rather than present a monolithic representation of the film sound professional across the American film industry, what is revealed is that the three major film communities had similarities as well as predictable contrasts that were developed as a result of the cultural differences.

The San Francisco Bay Area film scene is most often associated with George Lucas and his Skywalker Ranch. Certainly, when the average filmgoer, the film sound fan, and even the sound studies scholar are reflecting upon George Lucas, thoughts of Lucasfilm and Skywalker are foremost in one's mind. However, while Lucas located his empire in Marin County, he is not from the Bay Area, nor had he been involved with the local film scene in San Francisco. Lucas began his business in Marin County in tandem with Francis Ford Coppola's settling in San Francisco with American Zoetrope. To equate Lucas with the Bay Area and its film culture is to entirely miss what is crucial about many others who comprise the history and culture of Bay Area postproduction sound. While George Lucas and Skywalker Ranch are legendary for contributing excellence and innovation to film sound, for the purpose of revealing the unique nature of the San Francisco Bay Area filmmaking culture and its particular entry into digital film sound, the Lucas story is less illustrative. Readers curious about Lucas enjoy a plethora of information found in other sources such as Peter Biskind's *Easy Riders Raging Bulls: How the Sex-Drugs-And Rock 'N Roll Generation Saved Hollywood*,[41] *George Lucas: The Creative Impulse*, by Charles Champlin,[42] or, for a detailed accounting of the technological ventures of the Lucas empire, one can read *Droidmaker: George Lucas and the Digital Revolution*, by Michael Rubin.[43]

Instead, it is preferable to examine the sound community that revolves around the Saul Zaentz Film Center and Fantasy Records. Most of the Bay Area sound professionals worked at Zaentz and were closely linked to the goings-on in San Francisco. Many of these filmmakers came from the local documentary, art film, and public television scene and entered into feature films during the "all hands on deck" experience of *Apocalypse Now*.[44] Coppola's American Zoetrope proffered more opportunities for local Bay Area film neophytes than anyone up to that date, and it is this film that altered the artistic terrain of the Bay Area. Additionally, Saul Zaentz established himself as a producer of first-rate films such as *One Flew Over the Cuckoo's Nest* (1975), and his film legacy is one of deeply artistic and complex films, employing local talent at Zaentz's postproduction sound facility in Berkeley. Rather than rely on Lucas and his entrepreneurial endeavors to primarily define the Bay Area film culture, it is illuminating to look to Coppola and Zaentz to illustrate the burgeoning film scene that began in the 1970s, developed in the 1980s, and stabilized in the 1990s to help define the authentic Bay Area postproduction film culture. The Zaentz Film Center, which was centered in Berkeley, illustrates the transition from analog to digital in postproduction sound editing.

Hollywood's studio tradition provided an insular culture of the famous directors, producers, and film stars that defined American mainstream cinema. The industrialized nature of filmmaking hails from the beginnings of the studio system, which has resulted in a culture unlike any other.[45] Since the 1970s, corporations that are unaffiliated with filmmaking have bought studios for profitmaking and altered the terrain in Hollywood. However, while the economics have evolved over the past four decades, the celebrity culture that defines Hollywood and attracts investors has not.[46]

Postproduction sound has traditionally been considered a studio union job in Hollywood, but in the 1970s and 1980s independent boutique sound houses employing staff and freelance sound professionals began to compete with studio sound departments for postproduction sound budgets.[47] This was a peculiar time in Hollywood that allowed independent sound houses to develop innovative ideas in sound design and purchase equipment that was either owned by the small sound shop or the independent sound editor. By the 1990s, these sound houses were decked out with small mixing rooms, Foley and ADR stages, and at times small screening rooms. The popularity of the "big sound" films in exhibition provided ample opportunities for sound professionals to work long schedules and accrue overtime for their pensions and health benefits.[48]

Several postproduction sound companies had developed nascent digital systems for television shows in the 1980s. Television shows aired weekly and any methods that allowed faster postproduction workflows were an economic and creative asset for the predictable time crunch. These early technologies fueled the desire for more highly developed technological tools that would enable the television sound editors to produce their sound effects tracks quickly.[49]

To examine the transition from analog to digital sound editing in feature films in Hollywood is a more opaque process than in either New York or the Bay Area, as it is a behemoth. However, the 1990s was a decade when the studios were becoming more corporatized than ever before and the conglomeration of Hollywood put into focus a few key studios that controlled the majority of major film releases. Of these, only a few ventured into the digital editing early.[50] Sony bought Columbia Studios and was interested in mining the possibilities of the new digital technology.[51]

New York has a tradition of making director-driven films that engender smaller and more loyal crews. The New York film scene revolves around the directors who live and work there. These directors, Martin Scorsese, Jonathan Demme, Ethan and Joel Coen, Mike Nichols, and Woody Allen, among others, had worked with the same sound editors on each film.[52] The postproduction sound, until the 1990s, was primarily recorded and edited at Sound One, Hastings Sound, or Trans Audio Sound. Many of the same professionals worked at all three sound houses.[53]

New York films typically had a cultural signature: character-driven stories that were grittier and had simpler soundscapes. The sound editing duties ensured the dialogue was audible and that all necessary sounds were covered.[54] The 1980s brought some minor changes to this New York film sound sensibility when Hollywood sound editor Frank Warner supervised the sound on *Raging Bull* (1980), which had a New York picture editing crew, but films in New York usually were, and most often still are, about characters and story.[55]

During the 1980s, Hollywood was releasing films that required loud and expressive soundtracks and New York films were smaller, more intellectual, and often literary. The transition in New York sound design has its roots in the late 1980s with two New York sound professionals who are cultural New Yorkers, yet desired to break out of the monolithic and constrained traditions of the more naturalistic soundtrack. Both men were musicians, and Hollywood's entry into blockbusting soundtracks inspired one of them, Eugene Gearty, to spend a few years on the West Coast and learn some new techniques while soaking up some sun.[56] When he returned, he and his colleague, sound editor Skip Lievsay, greatly altered sound editing in New York, while retaining the integrity of the New York film narrative. Decidedly unsatisfied with the status quo of the rerecording mixer-driven soundtracks they had produced—soundtracks that relied on the discretion of one mixer with less input from the sound editors—they opted to open their own company, and made the initial leap into digital sound editing.[57] Before Hollywood or the San Francisco Bay Area were inclined to invest in any new electronic technology for feature films, New York film sound was well on its way in the new electronic transition. By 1991, although the workflow was not systematic yet, it was fairly predictable, and Lievsay had Ethan and Joel Coen as solid clients for whom he could create the soundtracks he imagined on technology he could control.[58]

Transition through Industrial Ethnography

This examination of the entry into digital sound editing is through a cultural geographic perspective. A technological discussion will focus on the tools, an historical discussion will delineate chronological development, and an aesthetic discussion will most typically include sound designers whose names are well known, or who have granted interviews. Until now, there has not been a specific study of the three principal American film cultures—Hollywood, the San Francisco Bay Area, and New York—as culturally different in their approach to sound design, nor in their reasons for the digital technologies they chose.

To date, few specific journal publications focus on sound editing as an art or craft, and then the assumption is that the reader is interested only in the celebrity sound designers or the latest technologies. One key exception is

Vincent LoBrutto's *Sound-On-Film*,[59] which includes excellent interviews with sound professionals in editing, mixing, and production mixing. Published in 1994, amidst this digital transition, LoBrutto captures some key insights that illuminate the process of a cross-section of sound "talent" and his contribution should not be underestimated. What is missing in this canonical text is the sense of cultural difference that is critical to a better understanding of the approach of the specific professionals and the effect they had on the continued development of digital sound editing.

Those who work in postproduction sound are almost invisible to those who are better paid and more influential and who also make the most critical decisions regarding a specific film production. The frustration stemming from this perceived marginalization comes into play in many ways, including economic concerns and industrial status, particularly within the Hollywood filmmaking process. Cultural examinations of Hollywood focus most often on the more visible aspects of filmmaking. Two seminal works, *Hollywood, the Dream Factory* by Hortense Powdermaker[60] and *Hollywood: The Movie Colony, the Movie Makers* by Leo Rosten,[61] introduce some of the culture of Hollywood and the film industry in the era, but Powdermaker and Rosten advanced different observations, which were informed by their different backgrounds. First, Rosten was a sociologist who had attempted screenwriting in Hollywood, so he had a bit of an insider's view. Powdermaker, on the other hand, was an anthropologist with no ties to the industry. Rosten had presented the first sociological study of Hollywood. His perspective was more of an industrial accounting, not unlike a study of any other powerful industry. Powdermaker reveals a more individuated and objective view of the subjects she encounters. My examination of this transition from analog to digital sound editing derives its cultural studies lens from my decades in postproduction sound, as one who is "from" the culture but no longer "of" the culture. Thus, my past experiences, my understanding of the culture as a sound professional, and my observations and research as a film sound scholar lead to a method of industrial ethnography of film sound, based on the deeply embedded knowledge of the workflows, the various roles of workers, and experience with the transition to digital editing myself. While industrial ethnography is more typically utilized by business to study the workplace, I adapt this method to study an industry with which

I have been deeply involved, but now view from the distance of the scholar. The danger of being too close to the subject of the study is one may fall into assumptions made prior to investigation. The tension between studying the object in film studies and making the object in film production allows bridges to be built if care is taken to critically assess assumptions from both views—not an easy task but can reveal additional perspectives.

The digital transition was not ubiquitous and omniscient, nor was it applied evenly across the American film industry. It was uneven, problematic, exciting, frightening, illuminating, and it empowered sound professionals while it also intimidated them.[62] The three film communities examined entered the transition in ways both similar and different. What is interesting is *how* and *why* the individuals involved made the choices that effected the trajectory of the transition in each of the three film cultures. This type of exploration requires ethnographic study, an understanding of the aesthetics and technology of sound design, and cultural experience with the individuals to be interviewed for deeper investigation.

Most postproduction professionals would disagree with the argument of technological determinism, "that certain supposedly unstoppable technical capabilities of computers will necessarily determine their future use,"[63] and it is also countered by Edward Friedman in his book, *Electric Dreams: Computers in American Culture* (2005), most notably for its disregard for human agency. From my ethnographic observations, this variable—that people adapt to technological tools, and *they* then adapt those tools to their own needs—holds true for the analog to digital sound editing transition. Friedman makes the point that history is full of stories of failed technologies that did not fulfill the needs or desires of the targeted consumer.[64]

Ruth Schwartz Cowan's "consumption junction," which details how and when consumers decide between competing technologies—that users adapt the technology to their needs, that they choose the technology they want, and that they do not respond to technologies that are ill-fitted to their needs[65]—is essential to understanding the transition to digital sound editing. The Social Construction of Technology (SCOT) theory claims that a technological artifact is not a linear construct, as determinism would argue, but is "multidirectional," and that it is important to view all stages of a technology as well as the "successful" ones.⁶⁶ The "winning" technologies that are the standard bearers today were developed in concert with the sound professionals who used them. Rather than simply conform to the technologies that were presented, the professionals utilized "work-arounds." As sound editors continued to adapt the tools, the more responsive technology developers repurposed their workstations to better conform to the needs and desires of the editors, and later the mixers who relied on the tools. Much as the latest generation of smartphones delete or alter features that are not pleasing the consumer, the editing systems are still modified and improved to satisfy the sound professionals.

A Writer, a Vampire, and an Explorer

I present three films as case studies as illustrations of the varying cultural approaches to the initial electronic sound editing transition. Prior to investigating these films, Chapter 2, "Geographical Cultures and Technological Tendencies," introduces the three film postproduction sound communities, and includes some relevant history and insight into the key players. I then lay out the development of the relevant technologies that were utilized in the inchoate electronic sound editing era of the 1980s and early 1990s. Each of the communities had different concerns and particular reasons for selecting the specific technologies in this initial transition.

The following three chapters examine one feature film from each community that was edited on these machines in the embryonic stages of electronic sound editing. In each of these films, I explore the principals involved, their introduction to the technology they used, how the work process was affected by the technological transition, and some of the aesthetic choices that were made for the sound design. I then present a textual analysis that I derive from some of the sound design the interviewees discuss. While the sound designers discuss their designs—some as individuals and some as collaborators—the analysis I present is my own interpretation of the aural narrative meaning. I include a flowchart for each film that depicts the use of technology in each film in the appendix.

Chapter 3, entitled "'Viscous Was the Word of the Day': The Interiority of *Barton Fink*," regards the work of Skip Lievsay, who engaged with the new digital technology with enthusiasm. Lievsay and several of his postproduction sound colleagues began their own company, c5, for the purpose of working with the new hard drive system Post Pro—developed by New England Digital Corporation—which had been responsible for Lievsay's favorite Synclavier synthesizer. Their reason for careening toward the inevitable electronic editing workflow was artistic: the desire to have boundless creativity, using less clumsy tools and to be able to control the workflow, including—eventually—the rerecording mix, was an attractive possibility that both professionals, especially Lievsay, did not want to postpone.

Lievsay has an almost symbiotic creative relationship with the Coens. One might have trouble discerning where one's creative ideas end and another's begin. With sound ideas actually scripted into the screenplay, and the both of the Coen brothers actively involved with brainstorming ideas with Lievsay and composer Carter Burwell, it becomes clear, after examining the construction of the soundtrack to *Barton Fink* (1991), that the four men author the soundtrack as a team.[67]

Chapter 4, "'How Would You Like to Work on a Monster Movie?': The Expressionism of *Bram Stoker's Dracula*," presents a unique opportunity for exploration. During the entire postproduction sound process of *Bram Stoker's Dracula* (1992) (*BSD*), I visited the Sony Studios lot and observed the editors and mixers, listened to the discussions that went on about the production, and was married to David Stone, the supervising sound editor. Additionally, I had been a Foley artist at the Coppola estate previously and performed that Foley in the very room where the final mix for *BSD* was executed. I had worked with Coppola's archivist, researcher, and, more recently, coproducer, Anahid Nazarian, previously, as well as with some of the sound crew.

When Coppola directed *BSD*, he intended to utilize new technology, but with the necessary financial support from the new studio giant, Sony. Unfortunately for Coppola and his postproduction sound professionals, they were snared in the web of budget tightening after the extravagance of Jon Peters and Peter Guber.[68] What this chapter reveals is how supervising sound editor David Stone and his talented crew met the economic and technological obstacles that continually disrupted the technologically challenged process with innovation, commitment, and collaboration, between atypical colleagues.

Chapter 5, "'The Sound of the Desert Is Tape Hiss': A Study in Contrasts in *The English Patient*," focuses on the film produced by Saul Zaentz and distributed by Miramax. This chapter will elucidate how Pat Jackson's small team of postproduction professionals managed to navigate through technological landmines while keeping Anthony Minghella's desire for pure aural reveries in the forefront. The intimate nature of Bay Area filmmakers who know each other well, have worked together, are friends, and are loyal to the culture of the area influenced their interaction when decoding solutions to the myriad of technological challenges that arose during the postproduction of *The English Patient* (1996).

Walter Murch's selection of Jackson as supervising sound editor proved to be a savvy decision as she shepherded the soundtrack through the challenges of a production dialogue track that was recorded using a new digital Nagra that was not compatible with the postproduction sound editing system Sonic Solutions. Jackson and her colleagues also faced countless picture changes as Murch reedited the continuity of a film that had two distinct stories, as well as the sound design challenge of two correlating soundscapes within the overall narrative.

The three case studies—one from each film community—illustrate the beginnings of the digital editing transition as it entered the feature film area as an awkward and inelegant one. The sound editors faced complex sets of challenges attempting to interface technologies that did not interrelate. Yet, in each case, these artists, these craftspersons, these professionals created soundscapes that were artistic, creative, and unique. The tools were, in most cases, supreme obstacles to be overcome. The creativity and ingenuity of these artists, and their desire and ability to collaborate, generated three distinct soundtracks that were recognized with awards at the time and are still considered—two decades later—to be three unusual and memorable aural narratives.

The concluding Chapter 6, "Reassessing Sound Design as a Collective Endeavor," reasserts the key premise that the sound professional is a creative artist who utilizes the tools necessary to produce and craft the soundscape of a film. The collaborative execution of motion picture sound design—rather than the myth of the singular and unique individual as ideator, or the beneficent leadership of the director who conceptualizes sound design while sound professionals merely execute through expertise and tools—becomes the model put forth to illustrate the creative nature of sound design, and its reliance on multiple individuals and their artistry. The three case studies are briefly revisited, as each has exemplified the insights from interviewees, and has demonstrated aesthetics particular to the film culture and style of each region, thereby reaffirming the differing approaches to the technological transitions, and the cultural contrasts regarding workflows.[69]

To begin, we first look at some recent historical developments in film sound for all three film communities—Hollywood, the San Francisco Bay Area, and New York, as they developed a sound style and artistic cultures with differing technological expectations.

Notes

1. Murch received the Life Achievement Award from the Cinema Audio Society in 1994 for his contributions to the profession. I attended this event and noted various views Murch shared on the industry, and the artform of sound. He spoke about his approach to sound design, editing, and mixing.
2. For some general background on the three filmmakers, a good book to skim is *Easy Riders Raging Bulls: How the Sex-Drugs-And Rock 'N Roll Generation Saved Hollywood*, by Peter Biskind, 1st Touchstone ed. (New York: Simon & Schuster, 2011).
3. Francis Ford Coppola, *The Conversation* (Lions Gate, 1974).
4. There have been numerous books and articles that discuss the three and their dislike and/or rejection of the Hollywood community and its restrictive labor structures. One need only look to *Droidmaker: George Lucas and the Digital Revolution*, 1st ed. (Gainesville, FL: Triad, 2012), by former Lucasfilm employee Michael Rubin, or *Whom God Wishes to Destroy: Francis Coppola and the New Hollywood*, by Jon Lewis (Durham, NC: Duke University Press, 1997). Also, any interviews or articles with Lucas, Coppola, or Murch include some allusion to the mistrust inherent in each man's dealings with Hollywood executives and the codified labor practices. For both artistic and business reasons, each of the three has his own reasons for eschewing Hollywood.
5. The three have contributed varying degrees of influence depending on the years involved and the role each has played. Also, they have contributed more as individuals than as a team.

6 Murch discussed the use of the term in relation to both films at the abovementioned Cinema Audio Society Awards banquet when he won the Career Achievement Award in 1994. To pinpoint an exact moment when the term became codified is not useful. It evolved over time during his early relationship with Coppola.
7 Coppola discusses this issue in his introduction to Michael Ondaatje's conversation with Murch about *Apocalypse Now* in *The Conversations* (New York: Knopf, 2004), in which he discusses film and sound editing with Murch.
8 Sound rerecording mixer Mark Berger also recollects (in my interview with him) that Coppola gave him his first credit, on *The Godfather Part II* (1974) as sound montage assistant. Berger recalls that he and Murch had both been dubbed sound montage-ists at some point. The purpose of this remark was to illustrate the difficulty of labeling exactly what the sound professionals were doing within some codified category, since they worked together in a fluid and collaborative manner.
9 I am using industry jargon. This is a common reference to indicate that you put in a sound where it is obvious and predictable.
10 German composer Oskar Sala composed the actual music for *The Birds*, but it was Herrmann who conceptualized the use of electronic music, using an early electronic sampler with Remi Glassman for the sounds used for the birds themselves and the soundscape.
11 Alfred Hitchcock, *The Birds* (Universal Studios, 1963).
12 Paul Verhoeven, *Robocop* (MGM (Video and DVD), 2014).
13 This is the term now accepted for the sound specialist who creates special designed sounds that are specific in nature for limited aspects of a film using a synthesizer.
14 Industrial ethnography.
15 The term "prosumer" typically refers to equipment sold to regular consumers that are at a more professional capability than a basic level for the amateur. The intended purchaser might be a film student or beginning professional who does not have the financial means for the full range of equipment used for industrial standards. Typically, it appeals to the user who wishes to feel above the beginner for less monetary output and is successful as a marketing tool.
16 Sound professionals who worked as sound designers in this capacity are Frank Serafine, John Fasal, and Alan Howarth, among others.
17 Iatrou edits and Lee is a rerecording mixer. Iatrou is credited as supervising sound editor and Lee as sound designer, although they collaborated on *La La Land* (2016) and *First Man* (2018).
18 Not unlike the "Great Man" theory in history that attributes progress to a substantial event led by one man rather than several smaller events that are connected and reliant on the agencies of many individuals.

19 Several of my interviewees—Pat Jackson, Mark Berger, and David Stone—have taught at prestigious film schools but are rarely interviewed. All three have worked with impressive directors, have won awards, and are considered by their peers to be exceptional in their fields.

20 One case study is a film by the Coen brothers, who include specific sounds in their scripts. I discuss their involvement with the soundtrack in the chapter about *Barton Fink*. Oliver Stone included sound effect and Foley notes in his script for *Platoon* (1986), on which I worked as a Foley artist.

21 The term "for hire" is an industry peculiarity that allows a production to own the creative work executed by the employee for the benefit of the film. While the artist creates the work, he or she does not own it and cannot claim authorship for purposes of future remuneration.

22 I use the reference to the Fordist assembly line method, where goods are in an assembly line, and one worker is responsible for one piece of the product. This does not include the assumption that the goods are low-cost, or that workers are "factory" workers, but that they take this process and adapt it to the production of a film for efficiency. This need for efficiency leads to an erroneous conclusion that the product must be inferior.

23 Jonathan Gray and Derek Johnson, *A Companion to Media Authorship* (Malden, MA: Wiley Blackwell, 2013), 350.

24 Ibid.

25 The gender-specific term is deliberate as most postproduction sound professionals were, and still are, male.

26 Vanessa Theme Ament, *The Foley Grail: The Art of Performing Sound for Film, Games, and Animation* (Amsterdam: Focal Press/Elsevier, 2009), 166.x.

27 I allude, again, to the "Great Man Theory," as explained above.

28 Rick Altman, *Silent Film Sound* (New York: Columbia University Press, 2004).

29 Dave Black, "Percussive Arts Society," *PAS Hall of Fame: Murray Spivack*, n.d., https://www.pas.org/about/hall-of-fame/murray-spivack.

30 Ibid.

31 Film sound professionals are more inclined to know more of Spivack's history and contribution.

32 Peter Franklin, "*King Kong* and Film on Music: Out of the Fog," in *Film Music: Critical Approaches*, ed. Kevin J. Donnelly (Edinburgh: Edinburgh University Press, 2001), 88–102. For an alternative view, which reveals a more collaborative approach to the sound design of *King Kong*, I recommend the special feature on the DVD of *King Kong* entitled "Max and Murray" in which Spivack discusses some of the circumstances surrounding their unusual relationship.

33 Hesmondhalgh refers to the "activity of symbol-making" in the cultural industries. David Hesmondhalgh and Sarah Baker, *Creative Labour: Media Work in Three Cultural Industries*, 1st ed. (New York: Routledge, 2011), 9.
34 David Hesmondhalgh, *The Cultural Industries* (London: Sage, 2012), 36.
35 Hesmondhalgh and Baker, *Creative Labour*, 81–112.
36 Synchresis: a sound put to an image that does not normally match, thereby creating a new meaning. Michel Chion, *Audio-Vision: Sound on Screen*, trans. Claudia Gorbman (New York: Columbia University Press, 1994), 63–5.
37 Digital workstations allow sound professionals to flow between editing tasks and mixing tasks with relative fluidity. While fully equipped mixing stages provide tools not available on the typical editing workstation, most sound editors can premix their edited tracks prior to the final mix today. This was not the case in the early 1990s.
38 It is problematic to define Hollywood, as it is actually the Los Angeles area geographically, but the "idea" of Hollywood, figuratively. Hollywood conjures up an image, a style, an era, and a filmmaking system. All are true, but not singularly definitive. For my purposes, Hollywood is all of these elements, as well as a geographical place that is the County of Los Angeles, from the San Fernando Valley, to the "West Side" to Downtown Los Angeles, to the actual city of Hollywood, in addition to the adjacent towns of Santa Monica, Beverly Hills, and West Hollywood, the Beach Cities, and parts of Orange County. For any "Angeleno" (one who resides in Los Angeles County), Hollywood means the film industry. For anyone not familiar with Southern California, it is difficult to contextualize.
39 The New York film industry refers to New York City, parts of the Hudson Valley, parts of Queens, parts of Brooklyn, and parts of New Jersey.
40 The San Francisco Bay Area film industry encompasses primarily San Francisco, Marin County, and East Bay, and parts of South Bay.
41 Biskind, *Easy Riders Raging Bulls*.
42 Charles Champlin, *George Lucas: The Creative Impulse*, rev. upd. Su ed. (New York: Harry N. Abrams, 1997).
43 Rubin, *Droidmaker*.
44 During the filming of *Apocalypse Now*, many neophyte film professionals got their start with Coppola. Within the film industry, it is historic for its length of production time and amount of staff employed.
45 Thomas Schatz, *The Genius of the System: Hollywood Filmmaking in the Studio Era* (Minneapolis: University of Minnesota Press), 2010.
46 Justin Wyatt, *High Concept: Movies and Marketing in Hollywood* (Austin: University of Texas Press), 1994.

47 Many of us employed during this time witnessed the transitions firsthand or worked at both studios and small sound houses.
48 Industrial ethnography.
49 Professionals in both television and feature film during this period were privy to the conversations of colleagues expressing various opinions of what would make a "better mousetrap" to enable them to more efficiently and more artistically create their soundtracks under the heightened time pressures. Industrial ethnography.
50 Industrial ethnography.
51 Thomas Mc Carthy, Jr., Personal Interview, interview by Vanessa Theme Ament in Hollywood, November 15, 2012.
52 Thomas Fleischman, Personal Interview, interview by Vanessa Theme Ament by Skype in New York, June 23, 2013.
53 Lee Dichter, Personal Interview, interview by Vanessa Theme Ament by Phone in New York, June 17, 2014.
54 Lee Dichter, Personal Interview.
55 This was one of those rare cross-pollinations of West Coast/East Coast postproduction crews. Picture editorial was in New York under Scorsese's editor Thelma Schoonmaker but sound editorial was in Hollywood under Frank Warner. However, picture and sound editorial always work closely together in feature films, so there is always collaboration between the two.
56 Eugene Gearty, Personal Interview, interviewed by Vanessa Theme Ament by Skype, in Beaufort, SC, May 22, 2014.
57 Skip Lievsay, Personal Interview, interviewed by Vanessa Theme Ament by Phone, August 18, 2013.
58 Skip Lievsay, Personal Interview (2013).
59 Vincent LoBrutto, *Sound-On-Film: Interviews with Creators of Film Sound* (Westport, CT: Praeger, 1994).
60 Hortense Powdermaker, *Hollywood, the Dream Factory: An Anthropologist Looks at the Movie-Makers* (Boston: Little, Brown, 1950).
61 Leo Rosten, *Hollywood: The Movie Colony, the Movie Makers* (New York: Harcourt Brace, 1941).
62 This characterization comes from the many responses from sound professionals as the transition was occurring.
63 Ted Friedman, *Electric Dreams: Computers in American Culture* (New York: New York University Press, 2005), 3.
64 Ibid., 30.
65 For information about the "consumption junction," see "The Consumption Junction: A Proposal for Research Strategies in the Sociology of Technology" by Ruth Schwartz Cowan in *The Social Construction of Technological Systems: New*

Directions in the Sociology and History of Technology (Cambridge, MA: MIT Press, 2012), 261–4.

66 "The Social Construction of Facts and Artifacts: Or How the Sociology of Science and the Sociology of Technology Might Benefit Each Other" by Trevor J. Pinch and Wiebe E. Bijker in *The Social Construction of Technological Systems: New Directions in the Sociology and History of Technology* (Cambridge, MA: MIT Press, 2012), 26–30.

67 Foley and ADR are included in the collaboration and authorship, and production mixers are part of this collaboration, as they record and mix the production dialogue, which is the essential foundation for any soundtrack.

68 To really understand the context of Sony, Coppola, and the Peters/Gruber extravagance, it is useful to read Nancy Griffin and Kim Masters, *Hit and Run: How Jon Peters and Peter Guber Took Sony for a Ride in Hollywood* (New York: Simon & Schuster, 1997).

69 Additionally, I include my impressions regarding the interviews I conducted with the professionals referred to and what I learned about the cultures from these interviews.

2

Geographical Cultures and Technological Tendencies

The Fordist Hollywood System

The Hollywood film sound process has been well codified for over half a century. Within this solid systemization, there has been a stable progression of development. By discussing my particular area of expertise, Foley, I can better explain how one simple aspect of postproduction sound is affected with a technological advance.

In the 1980s, a typical high-budget film allowed two or three weeks for Foley to be recorded—the industry term is "mixed"—and then allowed some days for pick-up cues. A television show required very quick turnover. It was not unusual for a one-hour episode to be Foleyed in one day, sound edited and rerecorded the next day, and aired the following day. Some higher-profile shows were allowed a week from beginning of the sound editing and Foley process to the final mix. Additionally, Foley artists were commonly pressured to perform synchronization that was so "tight" that on very close schedules, the rerecording mixers could "hang it as a unit."[1]

This might indicate just how formulaic postproduction sound had become for the everyday television shows.[2] In the early 1980s, analog sound had progressed to allow the recording of four tracks of Foley on magnetic film at once for a film or television show. This enhanced Foley artists' ability to blend and combine sounds for the final soundtrack. The result was that these artists could use a multitude of props for more organic creativity, and change shoes more frequently for variance in walking characters. Also, they could begin to add creative sounds to edited sound effects, rather than rely strictly on sound libraries. Foley became more a companion to effects editors as the technology allowed the artists to create more sounds on the stage. Thus, the balance of what was demanded from editors and artists shifted. In television, this meant the various television show schedules could shift to allow a more blended relationship with those working on a Foley stage. The Foley artists might be working on two different television shows during the same session for the same production company while the sound editors were busy "editing in"[3] the sound effects that were more routine to meet the increasingly shortened schedules. In feature films, it meant that the Foley artists would be asked to create more designed effects on the stage to be included as part of the overall sound effects for the film's soundscape.[4]

By the late 1980s, 24-track machines[5] were the norm on many feature and television rerecording stages. However, while television sound editorial was working faster than ever, the opposite was true in feature films. The divisions between those who worked in feature films were even greater as the technology allowed more exhibition pleasure for the viewer. Rerecording mixers could now premix many clusters of sound effects, background effects, Foley, and dialogue tracks, and thus present a rich and detailed final mix. THX and Dolby refined the quality and specificity of recorded sound, allowing for more control over the clarity and definition of even the smallest sounds. The result was a pristine soundtrack with precise and defined sounds. Mixers could place sound in front speakers, side speakers, and back speakers; move sound from one speaker to another; and add that famous "low-end boom" that made the room shake. Sound in film got bigger, louder, and more expansive.[6] 1987's *Robocop* won the Special Achievement Award from the Academy of Motion Picture Arts and Sciences for its spectacular feats in film sound editing.[7] This award was given for technological advancements in this era of filmmaking and preceded the convention of a more traditional award for sound effects editing as a technical *and* aesthetic award.

What is notable about workflow practices for the normal Hollywood feature film soundtrack was the structure of the sound crews. There were one or more supervising sound editors, three or more sound editors, two or more dialogue editors, one or two music editors, several assistant editors, two Foley artists, two or more automated dialogue replacement (ADR) editors, and a picture editing crew headed by one or more editors and several layers of assistants, all of whom were working simultaneously at a rapid pace and making changes in the film until the final mix.

This was a golden age in feature postproduction sound. Sound editors belonged to International Alliance for Theatrical Stagehand Employees (IATSE). Local 776, for picture and sound editors, and production and studio recording mixers belonged to IATSE Local 695.[8] Professionals worked long hours, had many weeks of employment on a film, and were able to jump from one film to another to prepare yet another massively expensive and technically amazing soundtrack. The vitality and exuberance expressed in the soundtracks of the 1980s reveal an excitement resulting from technological advances and a rising need for more highly trained professionals to create and construct the ever-growing separate sections of sound effects.[9]

By the late 1980s, digital editing was in its nascent incarnation. Television sound editors were transitioning into the digital workflow while the feature film sound editorial professionals were hearing that they would be required to learn digital editing within a few years to remain employable. In an interview for *MovieSound Newsletter* in 1991, with Leo Chaloukian, then president of the Academy of Television Arts and Sciences, I asked Chaloukian what he thought about the future of postproduction sound with the introduction of electronic[10] editing. His response as a facility owner whose bread and butter was in postproduction sound was this: "Why do you think electronic editing is proceeding the way it is? Because television can't afford the budgets."[11] I also asked Chaloukian about George Lucas's most recent pronouncement at the American Cinema Editors' banquet that sound was more than half the movie and should be taken seriously. I wondered how that comported with this scenario of speeding up and budget cutting we were all envisioning in postproduction sound. "Of course, I agree with him 100%. But you're talking about a man who does high budget films with (seemingly) unlimited funds. It'll apply in his case, because he can afford to do it. But 95% of the producers out there can't afford that luxury."[12] What these two statements from Chaloukian illustrate is the division that existed between television and feature postproduction sound in the beginning of what we now see as the digital sound transition. Chaloukian was keenly aware of the dichotomy. The transition that had hit television so quickly in the mid-1980s in Hollywood took longer to infiltrate the feature film postproduction sound workflow.

I offer as a case study of the Hollywood sound community's transition to digital sound the film *Bram Stoker's Dracula*, which won the Academy Award for Best Sound Editing of 1992. What makes this film an even more essential piece of the puzzle regarding cultural geographical tensions is its director, Francis Ford Coppola, and his problematic relationship with Hollywood.[13] Additionally, the film utilized sound professionals from both Northern and Southern California, which added to its complication, while attempting to make the digital transition in its inchoate form. This film was caught up in the travesty of the excesses of the picture executives, Peters/Guber at Sony, when *Dracula* first began shooting and the resulting studio budget tightened just as the film went into postproduction.[14] It is a story of studio versus independent filmmaker, and of Hollywood sound professionals comporting with Bay Area sensibilities.

The San Francisco Bay Area

The San Francisco Bay Area has been considered from the film industry's start to be a boutique film community. It is not unusual to equate George Lucas and the Bay Area as essentially the same since Lucas located his media operations in Marin County, just north of San Francisco. However, it is important to distinguish George Lucas from his two colleagues, Francis Ford Coppola and Walter Murch, although all three were attracted to the Bay Area for the same reason: it was not Hollywood. The original American Zoetrope production company, which was their initial alliance, had unstable financial endeavors until Coppola agreed to direct *The Godfather* (1972), which filled the coffers and solidified his stature as a director. All three men had never strayed from their initial visions of eschewing the Hollywood industrial style of filmmaking.[15] One of the hallmarks of the Northern California approach to filmmaking, early on, was in the value of allowing more time for artistic creativity.[16] In these earlier decades of postproduction, sound editors working at Skywalker Sound, in particular, were often allowed more time for editorial (and sometimes mixing) than their Southern colleagues.[17] The execution of sound design had a codified method that was particular to Skywalker and contrasted from methods employed in both Hollywood and New York.[18]

George Lucas's Skywalker Sound in Marin County

The Skywalker facility—known to many as "the Ranch"[19]—was built beginning in 1978, and is owned by George Lucas in Nicasio, Marin County, California.[20] Skywalker is nestled on a large and bucolic piece of land and surrounded by wildlife. During the 1980s and 1990s, there were two rerecording stages where even famous rock bands would record.[21] Additionally, the Ranch had a large and well-appointed Foley stage, several editing rooms, and machine rooms equipped with transferring equipment, coding machines for analog editing, and analog recorders. There was a luxuriously large screening room and two beautifully appointed rooms for sitting comfortably and enjoying coffee, vending machine treats, or gazing out the large windows. One of these rooms was named "the Heidi room" for its similarity to a Swiss chalet. Additionally, there was a room for recording ADR.[22]

Each sound editor would have a private editing room, some with window access to view a lake. Since there was no easy access to eateries anywhere nearby, most everyone stayed on the property all day. Meals were available on site. As the Ranch became better known, other directors wanted to do their postproduction there while staying at the remote surroundings. Add to this the proximity to San Francisco, and the emphasis on quality sound, and it was understandable why a Skywalker final mix would be included in a budget. Skywalker's fame derives as much from its environment as from its product. Lucas employed (and still does employ) a first-rate team of sound professionals from the Bay Area at the Ranch.

Skywalker provided longer work schedules for its editors, mixers, and Foley artists than did its neighbor across the Bay, Saul Zaentz Film Center. Through his Hollywood connections, and the emphasis on big-sound films, Lucas could attract larger budgets for his sound teams than the more independent Zaentz.[23] The approach to sound design at Skywalker was unique to the Ranch: take the time required, design special sounds, and pay people well.[24] The addition of Lucas's own patented exhibition sound system, THX, necessitated long and thorough soundtracks. In essence, Lucas was and is a businessman first. He understood vertical and horizontal integration and incorporated it into his media empire.[25]

Given Skywalker's powerful postproduction sound history, it might be logical to conclude that it would be representative of the Bay Area cultural practices in general. However, here are three reasons to consider Skywalker a unique and particular case. First, George Lucas designed a specific setting for his budding empire that did not originate in the Bay Area's cultural identity. While he has utilized Bay Area talent, early on Lucas returned to his alma mater, USC, where he found the blossoming talents Ben Burtt, and later Gary Rydstom.[26] Second, Skywalker's work practices reflect Lucas's own design and leadership, and are illustrative of his convictions regarding film sound rather than the Bay Area's regional culture. Third, there has been adequate coverage of the Skywalker soundtracks and their impact on film sound history.[27] DVDs often include special features narrated by Ben Burtt, Randy Thom, and Gary Rydstrom.[28] The Saul Zaentz Film Center in Berkeley typifies the San Francisco Bay Area culture and is another, and arguably more organic, illustration of the cultural postproduction sound practices.

The Other Side of the Bay: Saul Zaentz Film Center and Fantasy Records

Across the bay from Skywalker in the 1990s was the Saul Zaentz Film Center in Berkeley, the Bay Area's "other" postproduction sound community, the home of Saul Zaentz Productions and Fantasy Records. While the building at Berkeley is still there, it is now Fantasy Studios, a video game studio, and until 2007 was Concord Music Group.[29] During the 1980s and 1990s, several major films and their postproduction soundtracks were produced there, including *The Right Stuff* (1983), *Amadeus* (1984), *Blue Velvet* (1986), *The Unbearable Lightness of Being* (1988), and *The English Patient* (1996). George Lucas and Skywalker Sound are better known, but Mark Berger, who had been the principal rerecording mixer at Zaentz, has been nominated for and won four Academy Awards for Best Sound.[30]

At Zaentz, the film culture was not one of Los Angeles film school transplants who decided to leave the studio system and design a different film business model. Indeed, to fully understand the Bay Area film culture, and its influence on postproduction sound, one must examine the postproduction professionals and culture at Zaentz, or as the locals refer to it, Fantasy.[31]

Saul Zaentz first got involved with San Francisco music. He, along with a group of investors, bought the record company Fantasy Records, which distributed jazz artists, such as Gerry Mulligan and Chet Baker, and began to produce such groups as Creedence Clearwater Revival.[32] As Zaentz began to move into film, he quickly learned that "having your own mix studio gave you control in the Bay Area,"[33] and that this was an astute way into the filmmaking business. All of the local filmmakers, mostly from avant-garde or documentary film backgrounds, would need postproduction for their films. The Zaentz Film Center began making its money by duplicating the sound recordings required for editing and mixing in the days of analog sound.[34] Every sound effect had to be duplicated for editing on a Moviola. In the 1970s and 1980s, this was where most independent postproduction houses made their money. On the smaller films, many of them independent or low-budget, salaries were relatively low and schedules short.[35] Most of Zaentz's sound professionals were San Francisco natives or longtime transplants.[36] As Saul Zaentz began producing films, he kept his postproduction in-house at his facility in Berkeley. The Film Center kept busy with local independent films and small low-budget films supervised by associated sound editors.

Walter Murch and Alan Splet[37] were two of the most valued sound designers at Zaentz. While Murch was more independent, Splet worked primarily at Zaentz's facility in Berkeley. He brought in David Lynch as a client (they had started working in film together back in Philadelphia)[38] and had collected several valuable clients. Between Murch, Splet, Leslie Shatz, Douglas Murray, Pat Jackson, and Jay Boekelheide, Zaentz Film Center had an excellent reputation for sound design. Additionally, Mark Berger was considered a premium mixer. The workflow was faster than Skywalker, but not as fast as the Fordist factory model of Hollywood. Bay Area natives who had worked on local television programming and documentaries brought a sense of cultural pride to Zaentz.[39] The Zaentz faction was tight-knit and featured crossover practices between editing and mixing, since only one union represented everyone in the Bay Area film community. It was not unusual for someone to edit on one film and mix on another. Splet and Murch would edit and mix on their films.[40] Berger preferred to stay with mixing, but as the dialogue mixer, he was always the main mixer on the stage, unless Murch or Splet was on the show and mixing.

Roy Segal, the general manager of Fantasy and Zaentz Film Center, had been a recording engineer and studio manager at the Wally Heider Studios in San Francisco for such musicians as The Grateful Dead and Janis Joplin.[41] His expertise as both a facility manager and music engineer provided Segal revered gravitas for such a position. His methodology for management included having the music engineers for Fantasy Records learn the film mixing side of the company by training as Foley and ADR mixers. By the time a mixer was in a main mixing chair, he (most often, but sometimes she) knew film and music intimately. The lack of work boundaries—mostly due to economic necessity—allowed the professionals to learn skills that would otherwise be unavailable to them.[42]

As an illustration of a Saul Zaentz soundtrack to examine, both for aesthetics and for workflow, *The English Patient* is offered as a case study. The film won the Academy Award for Best Sound of 1996 and offers a fair glimpse of the initial transition from analog to digital sound, as well as of the postproduction sound work practices specific to the Bay Area sound culture. This film also demonstrates the increasing development of digital editing, as more film professionals were utilizing Pro Tools by this time and were working both in the northern and southern part of the state. The awkward technological transitions that plagued feature film sound continued to the end of the decade, but were evolving by this time, and Pro Tools was becoming the dominant editorial choice for the majority of Hollywood and New York sound editors.[43]

The Bay Area Film Industry Expansion

Until Francis Ford Coppola made the decision to make a film that resulted in bringing all of the potential film professionals in the Bay Area together on one film project, the media workers were focused on documentaries, experimental films, and the local public television programming.[44] Most, if not all, of the Bay Area film community was united for the length of the project. The film was an enormous undertaking, requiring massive changes, and was filmed on location. It required every professional within reach of San Francisco. It was an epic film and appears on the dossier of many postproduction sound professionals in the Bay Area today. The film was *Apocalypse Now* (1979). The film provided a start in sound for nearly everyone in the Bay Area with any desire to be in the industry.[45] Francis Ford Coppola and Walter Murch provided an ideal training ground for filmmakers. The three film students from Los Angeles—none of them natives of the Bay Area—have come to typify what most regard as the standard of Bay Area filmmaking. Yet it is important to remember the region's native film scene as it existed before they arrived. It is also critical that George Lucas, while close friends to both Coppola and Murch, has kept his business separate from their more "art-centered" filmmaking processes. Lucas has created what can be considered an empire around filmmaking,[46] while Murch and Coppola are more often singularly artists and filmmakers. Murch and Coppola have collaborated more often together than either has with Lucas, though both have also worked separately. San Francisco had one union that postproduction film professionals could join: IATSE and Moving Pictures Technicians Local 16. This union was neither terribly active nor very strong. Skywalker's sound professionals also had a different situation than those who worked for Saul Zaentz Film Center: a better salary and benefit structure. Lucas had bargained for better scale rates for his editors and mixers and this caused some divisions between the Ranch and Fantasy. Additionally, when the Bay Area sound editors and mixers merged with those in Hollywood Local 776, the Skywalker employees were allowed to take their pension contributions and transfer them to the new union affiliation, while the Zaentz employees were not.[47] This is another illustration of the differences between the well-connected Lucas and the more independent producer Saul Zaentz.

New York, New York

The New York film culture arose from a community of independent filmmakers beginning in the 1940s and included Arthur Penn, Sidney Lumet, Mike Nichols, Jonathan Demme, Woody Allen, Martin Scorsese, John Sayles, and, finally, Joel and Ethan Coen. One contribution to the health of the New York film scene was, interestingly, the Hollywood Blacklist of the 1940s and 1950s. The filmmakers who could no longer make a living in Hollywood went to New York to write for the thriving theatre world and eventually found themselves working in a film industry that was blossoming in New York.[48]

New York's theatrical and literary traditions bring a different lens to New York filmmaking. Stories that allow a "lack of closure" are more acceptable in the New York film culture. New York is not a "company town" and those who work in the film industry there experience more heterogeneity.[49] Independent producer James Schamus articulates the "three or four strands of production which New York still retains the archaeological traces of" as B-movie production, the contribution of the National Endowment of the Arts, the independent distribution companies—particularly New Line and Miramax—and the electronic media.[50] Actress Frances McDormand expresses the uniqueness of the New York film culture in contrast to Hollywood as "a different set of ethics ... you have to live differently in New York ... you don't have to hang out with people who do the same thing here. In fact, you're forced to hang out with people who don't, every day. So you're jarred out of complacency ... you can't fall back on clichés."[51] This sense that New York is more "real" and "connected" resonates with the New York sound professionals as well.

Beginning in the 1940s, Dick Vorisek, then the rerecording mixer at Reeves Sound Studio, was mixing the sound for New York-based directors. According to rerecording mixer Tom Fleischman, Vorisek pioneered New York film sound.[52] Vorisek would write down every fader position and every nuanced change in the equalization of a cue. Fleischman credits Vorisek with orienting the conditions that rerecording mixers continue to follow today in New York—most notably, the one mixer system.[53]

In the 1970s Elisha Birnbaum, a production sound mixer and Foley artist, left his native Israel to recreate his career in New York. At 24 years old, he rented a room in the famous Brill Building. Birnbaum remembers hiring a young man, Fleischman, to catalog Birnbaum's sound effects into a bona fide sound library. Years later, that young man would be one of Birnbaum's prize rerecording mixers at Sound One, New York's primary postproduction sound facility that Birnbaum invested in and took public along with partners Val Peters, Guy Spear, and Phil Pearl.[54] Sound One was a complete sound facility including editorial, Foley, music, and mixing. For over twenty years, Sound One was the major sound house for all of the notable New York filmmakers.[55]

Birnbaum is credited for training and/or employing all of the editors and mixers whose names now appear in most New York sound mixes. He is the foundational source for the postproduction sound community for New York. Birnbaum's sensibility as a production mixer influenced the New York style of sound—which is determinedly more realistic and less "hyperrealistic" than Hollywood. As Birnbaum states, "I wanted to do realistic sound." Before turning the Foley reins over to others, Birnbaum was the primary Foley artist and has forever influenced the style of Foley in New York. "I was the only recording engineer from location who did Foley and I knew how it sounded."[56]

New York had several other postproduction sound facilities that offered some competition to Sound One. Recording engineer Emil Neroda owned The Sound Shop, Dan Sable owned Hastings Sound, and Dick Vorisek, along with his brother Jack, headed Trans Audio. However, New York was and still is a small and tightly knit film community, and the postproduction sound professionals who live and work there moved within their small professional community. Mixers belonged to IATSE Local 52, while the editors belonged to IATSE Local 771.[57] In 1980, William Bender, the business agent for Local 771, negotiated the "Collaborating Editor Clause" to ensure that union New York editors were hired on standby status to work on any film production from another jurisdiction that came to New York. Additionally, in 1981, Bender negotiated a Reciprocity Agreement with Hollywood's Local 776, which was the first step toward a merger into the national editors union, Local 700 in 1998.[58] In 1999, the film studio mixers who belonged to Local 52 merged with Local 770 as well.[59] Those who were production mixers remained in Local 52.

In 1989, Skip Lievsay, Bruce Pross, Phil Stockton, and Ron Bochar decided to break away and start their own company, which they named c5.[60] The idea was simple: by building a Foley stage that Bochar could perform on and Pross could mix on using the new Digital Workstation Post Pro, recently developed by the New England Digital Corporation, they could finance their own company and have more control over their own product. Eventually, they hoped to mix their soundtracks digitally, but in the beginning, they knew they would have to continue to mix at Sound One.[61] Lievsay had acquired his own artistic and cultural capital in New York as the sound designer for the Coen brothers. His entry into the Coens' film *Blood Simple* (1984) solidified Lievsay's New York reputation, although he was not a known entity in Hollywood until after his Golden Reel-winning soundtrack for *Barton Fink* (1991), which is my case study of the New York entry into digital sound editing. The Palme d'Or-winning film, with its New York sensibility and offbeat visual and aural setting, put the Coens, Lievsay, and composer Carter Burwell on a solid trajectory of control over their respective film careers that has continued.

The Technological Geography: A Journey from Coast to Coast

The purpose of the technological discussion here is to clarify the distinctions, as well as collaborations, that occurred in the mid-to-late 1980s and into the early 1990s that contributed to the cultural differences between the three film communities in discussion. What follows is the historical context of that era that more completely clarifies the development of the early beginnings of the digital audio workstation (DAW) that then plays a critical character in the differing roles within the cultures of New York, Hollywood, and the Bay Area, as they made the transition in postproduction sound to editing digitally.

My purpose is not to focus on how the digital experimentation finally led to Pro Tools as the contemporary accepted standard in mainstream American postproduction sound editing and mixing practices. Nor is it to include a trajectory that involves the development of digital rerecording mixing, mastering, and exhibition. Rather, it reveals the direction each of the three film communities—New York, Hollywood, and the San Francisco Bay Area—took when making the critical decisions to begin the transition to digital sound editing in major motion pictures. Both New York and Hollywood had been incorporating some digital sound editing into television and low-budget filmmaking since the early 1980s. What seemed as an inevitability for the major film production business—to transition to digital sound editing—actually was awkward and unpredictable, much as the transition from silent film to early sound pictures had been.

Film Follows Television

In Hollywood, the divisions between those who worked in postproduction sound in major films and those who worked in television were divided by location, status, and even industry academies.[62] The work practices, even though overseen by the same union rules, follow different schedules and different standards.[63]

In the late 1970s, at Glen Glenn Sound, Emory Cohen, vice president of operations, commissioned engineers Jim Fulmis of Fultek and Skip Holt to design a system that allowed dubbing sound to video for television shows using SMPTE time code. The goal was to add sound to film for the television shows that needed minor sound effects and some sweetening. Sound effects were kept in a video library and synchronized into the mixed tracks by the rerecording mixers. This system was called Post Audio Production (PAP) and a select group of rerecording mixers became specialists with the system.

In 1989, Cohen, by this time the president of Pacific Video and its postproduction sound facility, Laser Pacific, was developing a digital editing system that mimicked the analog editing bench visually. Cybermation, renamed WaveFrame, developed the DAW to be used by IBM compatible computers. Chuck Grindstaff, an aerospace signal processing engineer, along with his father Doug Grindstaff, a veteran sound editor, collaborated on the endeavor. Grindstaff the elder was hired to help promote the new system and encourage studios and independent sound facilities to implement this new digital system.[64] Grindstaff was well respected and had been in the television end of postproduction sound for twenty-five years.

The Hollywood editors' guild, at that time Local 776, offered free classes to all union editors to learn "Cyberframe."[65] Coincidentally, the primary teacher of these classes, Richard Steven, was a supervising sound editor at Laser Pacific.[66] It was not long until most of the studios with any television shows on the lots were invested in the WaveFrame digital system. Feature films were still having their postproduction sound edited on magnetic film. The division between television and feature films kept digital editing from infiltrating major motion pictures for a few more years. Even so, most sound editors were anxious about the oncoming "digital revolution" and what it might mean for their employment futures.

Soon after these Hollywood industry professionals ignited major interest in this manageable and user-friendly system, a larger corporation, Digital F/X in Mountain View, California, acquired WaveFrame.[67] The belief that WaveFrame would become the industry standard was widespread. This acquisition, announced in September of 1992, occurred during the postproduction sound editing and mixing of *Bram Stoker's Dracula*, which was the first major film from Hollywood that employed the WaveFrame technology as the primary editing platform. Brief mention should be made of another contender for the successful DAW: the Fairlight Electric Sound and Picture (ESP) system, which was a reconfigured electronic editing system based on the Fairlight synthesizer. In 1989 the Fairlight ESP was adapted by Warner Brothers and Todd AO and used in editing, rerecording, and, later on, in Foley mixing.[68] The Fairlight was a hybrid analog and digital video processor that was more expensive and did not interchange with other electronic systems, such as the AVID. It was an excellent recording machine and was superior for "punch-in" recordings on the Foley stage.[69] However, it only accessed up to twenty-four tracks, which was already available in analog recording machines that had been previously installed in many mixing stages. The initial outlay of expense to include the Fairlight, and its inability to interface with other developing technologies, caused it to become obsolete in Hollywood. Most facilities were not willing to make the financial investment necessary for a system that was not yet fully adaptable.

DroidWorks Didn't

It would stand to reason that George Lucas would get into the mix regarding the development of a digital editing system. The nonlinear picture editing system, called EditDroid, made its debut in 1984 at the National Organization of Broadcasters, in Las Vegas.⁷⁰ It was a joint venture between Lucas's spin-off corporation DroidWorks and Convergence Corporation. EditDroid was a laser disc-based system and never enjoyed commercial success. In 1992, Pat Jackson, a Bay Area sound editor, remembers that the *Young Indiana Jones* television series was edited on EditDroid.⁷¹ The software was sold to Avid Technologies in 1993.⁷²

Lucas also put his team to work on SoundDroid, a digital sound editing system, developed by James A. (Andy) Moorer.⁷³ SoundDroid was a hard disk system that utilized an audio signal processor. Only one prototype was constructed between the years of 1980 and 1987. Sprocket Systems, Lucas's original sound editing company, transitioned into Skywalker Ranch in 1987 and moved from San Anselmo to Nicasio, both in Marin County.⁷⁴ SoundDroid was never commercialized, and Skywalker continued to provide its sound services using the same analog standards as both Hollywood and New York. However, Moorer developed a new digital editing platform called Sonic Solutions and along with his partners, Robert J. Doris and Mary C. Sauer, received the Emmy for Technical Achievement in 1996 for this contribution.⁷⁵

The original intention of Sonic Solutions was noise reduction. Mix Magazine featured a small article celebrating Sonic Solutions and its contribution to the digital sound revolution:

> NoNoise®, a Macintosh-based system that applies proprietary DSP algorithms that eliminate broadband background noise, as well as AC hum, HVAC buzz, camera whine and other ambient noises. NoNoise could also reduce overload distortion, acoustical click/pops, transients caused by bad splices and channel breakup from wireless mics—without affecting the original source material. Sonic Solutions eventually expanded into 2-channel and multichannel workstation development and developed the first DVD premastering system.⁷⁶

In 1988, Sonic Solutions expanded its capabilities to include mixing, equalization, and CD mastering. By 1995, sound editing was added with the Multitrack Sonic System, availing users to sixty-four tracks of input and output. Additionally, Sonic Solutions software was incorporated into Pro Tools and Avid.

Roy Segal, vice president and general manager of Fantasy Records and the Saul Zaentz Film Center in Berkeley, adopted Sonic Solutions for use at

Fantasy Records. Its noise-reduction abilities and its advantages in mastering were invaluable for Segal, who had a two-pronged company: Fantasy Records, which was recording and mastering CDs, and the Film Center, which was providing postproduction sound services on major films. According to Steve Shurtz, former operations manager of both facilities at the time, Segal initially used Sonic Solutions NoNoise for cleaning and mastering of CDs on the music side, Fantasy Records, and later incorporated it into the Film Center's projects as well.[77] The system was not in operation on the rerecording mixing stage, nor was it utilized for Foley mixing or for ADR.[78]

Segal wanted to keep the film transfer workers[79] of the Film Center working. Therefore, he did not want to change over to a digital platform until he was sure which one was the wisest investment. Segal asked the engineers their preference, which was Sonic Solutions for its noise-reduction qualities. None of the sound editors who worked at the Film Center were asked about their preferences. Many of these editors were also freelancers who had purchased their own equipment. So for uniformity, when they were employed on a Zaentz film, Segal required the editors to use Sonic Solutions.[80] The irony of this is the system was not as ideal for sound editing, since it was a more "piecemeal" system: the editor was required to move a finished sound effect to another track, and it did not work well for a complete track that required only minor editing and finessing. It was, however, a workable system for dialogue editors, who would take edited parts of the production track and put the cleaned and refined piece into the new edited track. Sound effects, background effects, and Foley editors found Sonic Solutions problematic for the practice of building a new effects track that required a more linear approach.[81]

While the San Francisco Bay Area film sound professionals were juggling their attentions between Sonic Solutions and magnetic film, Sound Tools, a new digital system was entering the scene in 1989. It began as a small disk-to-disk recording and editing system designed by Evan Brooks and Peter Gotcher, friends from the University of California at Berkeley.[82] They had started with a digital drum machine, the Drumulator, which launched their company Digidrums, later to be renamed Digidesign. Sound Tools, which their introduced in 1989 at the National Association of Music Merchants (NAMM) show, was the prototype for Pro Tools, the present-day industry standard. At the time, however, Sound Tools was only nominally capable of what major films would require from any editing or recording tools. Still, this new digital platform was gaining attention from fellow Bay Area sound professionals.[83] The third case study, which illustrates the San Francisco Bay Area's entry into digital sound editing, is *The English Patient*, which released in 1996. By the mid-1990s Pro Tools had become the industry standard in Hollywood and New York. However, Segal had remained loyal to Sonic Solutions, so to keep profits in-house, the editors were required to edit on Sonic Solutions when the rest of the industry had moved on.[84]

"The Bleeding Edge"

In 1980, the Academy of Motion Pictures Arts and Sciences awarded the Scientific and Engineering Award to Neiman-Tillar Associates and Mini-Micro Systems, Inc., for inventing the first digital sound editing system.[85] According to its inventor at Mini-Micro Systems, William R. Deitrick, this system had been in use since January of 1977.[86] It could locate and edit sound effects from a sound library that was stored on a hard disk computer, and was first used at Trans Audio Video (TAV), in Hollywood, for television shows and low-budget films.[87]

Eugene Gearty, a young and eager new sound professional from New York, moved to Los Angeles specifically to work on this system in 1983 at TAV. The system, Automated Computer Controlled Editing Sound System (ACCESS), was the first generation of what would become many iterations of digital editing systems—all vying to be the standard in the industry—for sound editing and eventually sound mixing for major motion pictures in Hollywood. Gearty, a former jazz musician and audio engineering student at Berklee College of Music in Boston, substantiates, "Larry [Neiman] and Jack [Tillar] were on the bleeding edge . . . There were up and comers all the time . . . It was a hotbed of ideas in the late '80s to be the first one with a digital workstation."[88]

Gearty had been excited to work with a digital-based system for sound design since his background was music. "I'm glad I didn't go to film school. I saw the process as one."[89] Gearty was not trained in the tradition of sound connected to narrative, so his perspective was that of sound as part of imagination and exploration. The prospect of working in Hollywood where big sound films were starting to be made and working with a tool that was more freeing than a mechanical film-based machine was reason enough for the native New Yorker Gearty to leave New York for a while. Additionally, his idols, including another musician turned film sound professional Alan Splet, were in California making imaginative soundtracks. The films being produced in New York at this time were narrative-based and not the genres that required the special sound tracks that Hollywood was producing.

Gearty had been working at a postproduction sound facility called the Sound Shop, owned by Emil Neroda, who was an experienced recording engineer with Reeves Recording Studio on 44th Street.[90] Gearty and Skip Lievsay, a bass player-turned sound editor,[91] had met and begun a friendship at Neroda's facility. This friendship would later become the seed for one of New York's major sound facilities in the 1990s, c5 Sound.

Once Gearty began his work at TAV, he gained experience editing on the ACCESS system and saved his money to buy his own Synclavier digital synthesizer, which had been his goal since he was first introduced to one at Berklee. The Synclavier was a keyboard random-access memory (RAM) system that allowed the user to create special design sounds.[92] Musicians had been using the Synclavier for years in recordings and in live performance.[93] For Gearty, this new tool provided him a method for designing sounds for the television shows and small films he was employed on.

Gearty remembers that his colleagues in Hollywood during his tenure there (1983–7) were anxious about the oncoming digital transition. He recalls that the general conversation was, "Let's get ourselves out of this slow system and get into something efficient."[94] The need to work quickly in television and make a profit was a driving force for the development of the DAW.

The New England Digital Corporation had developed the Synclavier, and in 1984, the Direct to Disk ("D to D") model was introduced. Gearty was working on the "D to D" Synclavier as well as the ACCESS in Hollywood. Neither system was a stand-alone workstation. That development was still to come. In 1987, two events occurred that propelled Gearty back to New York. First, Sound One acquired a Synclavier similar to what he had worked on in Hollywood. Second,

Neroda had added the second ACCESS system in the country to his sound arsenal at the Sound Shop. Gearty went back to New York and continued to build his résumé, as well as his skill set on the digital platforms of the two systems. He also encouraged Lievsay to transition from magnetic film sound editing to the Synclavier system at Sound One.

The "D to D" model of the Synclavier was the prototype for what would be New England Digital Corporation's Post Pro digital recording and editing system that both Gearty and Lievsay employed when they opened c5 Sound in New York in 1989.[95] Post Pro offered a stand-alone DAW and Lievsay opted to utilize it as he transitioned to digital editing for a major motion picture, before both San Francisco Bay Area and Hollywood.[96] My case study that discusses the digital transition in New York is *Barton Fink*, released in 1991.

Notes

1   "Hang it as a unit" refers to the practice of actually hanging the full coat of Foley (three tracks) as a complete unit, with no separate tracks edited. The mixers would mix the Foley in with the sound effects "as-is" and the sync would stay as it was performed.
2   I am emphasizing the process in television in opposition to the more time-intensive and careful planning that was practiced in feature film workflows.
3   "Editing in" is a term of art that refers to editing the sounds into the actual magnetic film track. Picture editors edit or "cut" what already exists. Sound editors put in what needs to be added.
4   For the film *Who Framed Roger Rabbit?* (1988) Foley artists John Roesch and Ellen Heuer created many of the animated and designed effects for sound supervisor Chuck Campbell on the Foley stage. Campbell did not have a sound effects library that contained the necessary components, and he had empowered the Foley stage to design effects. Campbell's reliance on the Foley stage had contributed to the stage becoming used more to supplement sound effects than is the original purpose of Foley. Campbell received the Academy Award for Sound Effects for the film. Many of us in postproduction sound believed the Foley artists contributed greatly to the work recognized in this award.
5   The 24-track machine was analog tape, rather than magnetic film, and ran on many rerecording stages, and by the mid-1990s, some Foley stages for recording Foley. The Foley was then duplicated onto the specific technological format (magnetic

film or digital) depending on what machines or computers editors were using. It was confusing, complex, and created solutions and problems.

6   Industrial ethnography.
7   Found on the Academy Awards Database for Archived Oscars: Stephen Flick, John Pospisil, http://awardsdatabase.oscars.org/search/results.
8   I explain the union affiliation because editors and mixers were in separate unions at the beginning of the digital transition, but by the late 1990s, all three cities had combined rerecording mixers and editors into one union for ease of workflow, transmitting sound tracks over the internet, and enabling workers to move back and forth between cities for film projects.
9   Industrial ethnography.
10 At the time, we did not refer to this new technology as digital. We called it electronic editing.
11 Vanessa Ament, "Leo Chaloukian Interview, Cont.," *MovieSound Newsletter* 2, no. 1 (February 1992): 3.
12 Ibid.
13 Jon Lewis and Eric Smoodin, eds., *Looking Past the Screen: Case Studies in American Film History and Method* (Durham, NC: Duke University Press, 2007).
14 Nancy Griffin and Kim Masters. *Hit and Run: How Jon Peters and Peter Guber Took Sony for a Ride in Hollywood* (New York: Simon & Schuster), 1997.
15 Until, of course, Lucas's recent sale of his Lucasfilm to Disney. Coppola's struggles with Hollywood are described in Jon Lewis, *Whom God Wishes to Destroy . . .: Francis Coppola and the New Hollywood* (Durham, NC: Duke University Press, 1997).
16 This is a cultural quality of the Bay Area that is appreciated by many. Taking appropriate time for sound has been a hallmark of Lucas.
17 Until all sound editors in the three cities were included in one union, some in the Bay Area assumed that Hollywood sound editors followed a previous convention of the studio days where sounds were formulaic and "pasted" in without design. Conversely, there were Hollywood sound professionals who wondered how Bay Area sound professionals would adapt to the fast pace of Hollywood practices. As the two regions became united in the same union, and standardized technologies afforded them the opportunity to work together, the misconceptions faded, and the mutual appreciation increased.
18 My experiences as a Foley artist provided me a unique perspective of the distinctions between the processes of both the Zaentz facility and Skywalker Ranch. While Zaentz and Skywalker did not operate under the same union paradigm as Hollywood or New York, the editing and Foley practices economically were more similar to Los Angeles or New York. Skywalker, in contrast, had different economic

realities that resulted in different work practices. One example of this is a particular practice at Skywalker during this era, where someone would be assigned to move the microphone for me for every cue, and write down every microphone setting, track assignment, surface changes, and character. This is unfamiliar to a Los Angeles–trained Foley artist as this practice is cost-prohibitive in the Los Angeles system.

19 Most industry professionals refer to the facility as either Skywalker or the Ranch.
20 The purchase of Lucasfilm by Walt Disney Studios did not include the actual Skywalker facility.
21 Personal observation of the Rolling Stones there during a 1989 recording session.
22 Industrial ethnography.
23 Malcolm Fife, Personal Interview, by Vanessa Theme Ament by Phone in San Francisco, CA, June 29, 2013.
24 For a better accounting of the Lucas Sound development and the more specific differences of the major influences of Lucas and his tremendous team of Burtt, Rydstrom, Thom, and the sound maven Walter Murch, I recommend Midge Costin's documentary *Making Waves* (2019).
25 This is a fair conclusion after reading *Droidmaker: George Lucas and the Digital Revolution*, 1st ed. (Gainesville, FL: Triad, 2012).
26 Randy Thom, who is a San Francisco native, approached Lucas for work on the set of *American Graffiti* (1973). Other USC hires are discussed in detail in *Droidmaker*.
27 Again, I refer to abovementioned titles as a beginning for Lucas research.
28 *Saving Private Ryan* (1998—Rydstrom), *The Incredibles* (2004—Thom), and *Wall-E* (2008—Burtt) are examples.
29 Concord Records acquired Fantasy and became Concord Music Group. In 2007 Wareham Property Group, owners of the Saul Zaentz Media Center, purchased the property and developed Fantasy Studios.
30 Rubin, *Droidmaker*.
31 Part of the local color of the Bay Area film culture is to call Zaentz Film Center "Fantasy." The beginnings of the Film Center were in the recording industry with the record label Fantasy Records, and all locals refer to both the film and the recording aspects of Zaentz's company as Fantasy. This is partially because the mixers often crossed over from working in the music side to the film side and back again. However, it is a local tradition.
32 Duane Byrge and Mike Barnes, "Legendary Producer Saul Zaentz Dies at 92," *Hollywood Reporter*, January 2, 2014, 1.
33 This is a quote from mixer Mark Berger, as he and I talked during an interview. Berger stated that it became common knowledge in San Francisco that the way to build your film business was to have a postproduction facility. This gave you

34 Mark Berger, Personal Interview. complete control. He noted that Coppola, Lucas, Murch, and Zaentz all knew this and followed this formula. Mark Berger, Personal Interview, interview by Vanessa Theme Ament, Skype, in Berkeley, CA, July 17, 2013.
34 Mark Berger, Personal Interview.
35 Malcolm Fife, Personal Interview (2013).
36 Ibid.
37 Had Alan Spelt not passed away in 1994 of cancer, it would be interesting to see what his contributions to the digital age would have been. Splet was an accomplished cellist prior to becoming a sound artist and was known for preferring fresh recordings and innovative sound techniques rather than rely on old tried and true sound effects and conventional approaches. Splet was revered by all three film sound communities as one of the most artistic and compelling sound designers, yet he died just short of his fifty-fifth birthday and was not one to involve himself in interviews. His wife, Ann Kroeber, has kept his library intact and discusses their work with Lynch and other directors freely. Ann Kroeber, Personal Interview, by Vanessa Theme Ament, July 24, 2013.
38 Ann Kroeber, Personal Interview.
39 Pat Jackson, Personal Interview, interview by Vanessa Theme Ament, Skype, in San Franciso, CA, July 18, 2013.
40 After Splet's death, in 1994, a second rerecording stage was built on the second floor of the Film Center called The Splet Stage.
41 Malcom Fife, Personal Interview, interview by Vanessa Theme Ament, Phone, in San Francisco, CA, June 1, 2014.
42 One such professional, Michael Semanick, hailed from Berklee College of Music, worked as both music engineer and Foley mixer on several projects at the Zaentz Film Center, and became an Academy Award–winning rerecording mixer on several films including *The Lord of the Rings: The Return of the King* (2003) and *King Kong* (2005).
43 This assertion will be amplified and explained in the chapter on *The English Patient*.
44 Personal interview with Pat Jackson with Vanessa Ament, by Skype, in San Francisco, July 18, 2013.
45 Ibid.
46 Lucas is quoted as saying in reference to Coppola, "He says he's too crazy and I'm not crazy enough. Francis spends every day jumping off a cliff and hoping he's going to land OK. My main interest is security." Rubin, *Droidmaker*, 10.
47 Malcolm Fife, Personal Interview (2013).
48 Tod Lippy, ed., *Projections 11: New York Fill-Makers on New York Film-Making* (London: Faber & Faber, 2000), introduction, x.
49 Ibid., ix.

50 Ibid., 5.
51 Ibid., 22.
52 Tom Fleischman, Personal Interview, interviewed by Vanessa Theme Ament, Skype, June 23, 2013.
53 Ibid.
54 Elisha Birnbaum, Personal Interview, interview by Vanessa Theme Ament by Phone in New York, September 22, 2013.
55 Staff, "Sound One Corp.," *Mix Magazine*, September 1, 2000, archived https://www.mixonline.com/.
56 Elisha Birnbaum, Personal Interview.
57 Thomas Fleischman, Personal Interview.
58 Louis Bertini, "The Busy Boom Years of Local 771," *Editors Guild Magazine* 1, no. 5 (October 2012): 1.
59 "Guild's History," *MPEG*, accessed June 15, 2014, archived https://www.editorsguild.com/.
60 Amy Taubin, "A Sound Is Built," *Village Voice* 34 (October 3, 1989): 71–2.
61 Skip Lievsay, Personal Interview, interview by Vanessa Theme Ament, August 18, 2013.
62 The Academy of Motion Picture Arts and Sciences is an honorary association for those who are deemed qualified and have worked in the film industry. The Academy of Television Arts and Sciences is the equivalent organization for those in the television industry. Many professionals belong to both, and some belong to only one or the other.
63 All union professionals were encouraged to be trained on the technologies for editing purposes by the editors' union, which was initially 776, and became the national union 700.
64 Leon Silverman, "The New Post Production Workflow: Today and Tomorrow," *Motion Kodak*, accessed May 30, 2014, 36.
65 Cyberframe and WaveFrame are the same system. The name fluctuated.
66 As a member of the union, I took the classes with Richard Steven, and later with Richard Corwin, another sound editor who had worked with Doug Grindstaff. This platform was considered the primary system for most Hollywood television shows in Hollywood in the early 1990s.
67 "Digital F/X Announces Acquisition of Waveframe; Merges into New Digital F/X Audio Digital F/X," The Free Library, *PR Newswire* (September 30, 1992).
68 Malcolm Fife, Personal Interview (2014).
69 Punching in is the action of playing the prerecorded Foley and clicking in the new recording in and out of the old recording. It is considered a sophisticated skill for a Foley mixer and the Fairlight had advantages for mixers in this execution.

70 *EditDroid NAB 2*, Video (Las Vegas, NV).
71 Pat Jackson, Personal Interview.
72 Rubin, *Droidmaker*.
73 James Moorer, "James A. Moorer Résumé."
74 I was privileged to see EditDroid demonstrated by Ben Burtt at Sprockets Systems in 1988, while Skywalker and Sprockets coexisted. Burtt told Dave Stone and me a story about how George Lucas wondered if there was going to be a SoundDroid. The response? "Yes, George. It is unworkable. It is in storage in San Rafael. Remember?"
75 "Sonic Solutions NoNOISE Honored with Emmy Award; Recognized for Groundbreaking Digital Sound Restoration Product," *Business Wire*, October 2, 1996.
76 "1978 New England Digital Synclavier," *Mix Magazine*, September 1, 2006.
77 Steve Shurtz, Personal Interview, interview by Vanessa Theme Ament, Facebook Private Message, May 29, 2014.
78 Mark Berger, Personal Interview.
79 These workers were responsible for transferring sound effects and Foley recording to magnetic film for editing purposes, and for rerecording mixes.
80 Pat Jackson, Personal Interview.
81 Eugene Gearty, Personal Interview.
82 "A Brief History of Pro Tools," musicradar.com, accessed May 30, 2014.
83 Steve Shurtz, Personal Interview.
84 Pat Jackson, Personal Interview.
85 "Academy Awards Database," *Oscars.org*, accessed May 30, 2014.
86 InfoWorld Media Group Inc, *InfoWorld* (InfoWorld Media Group, 1979).
87 Eugene Gearty, Personal Interview, interview by Vanessa Theme Ament, Skype, in Beaufort, SC, May 22, 2014.
88 Ibid.
89 Ibid.
90 Ibid.
91 Lippy, *Projections 11*, 40.12 p12.
92 "1978 New England Digital Synclavier," 1.
93 As a professional musician at the time, I watched many of my colleagues use the "Synclav" in live shows as well as in recording.
94 Eugene Gearty Personal Interview, 2014.
95 Ibid.
96 Skip Lievsay, Personal Interview, 2013.

3

"Viscous Was the Word of the Day": The Interiority of *Barton Fink*

> The Coens use sound "as a whole 'nother character" in the film.[1]
> —Lee Dichter, rerecording mixer

Academy Award winner Skip Lievsay (*Gravity*, 2013) (Plate 1) worked with the Coen brothers[2] to develop a unique sense of interiority in *Barton Fink* (1991), which relies heavily on sound design to create the uncanny environment of Barton's hotel, and most notably his room. Rather than rely on a nondiegetic film score to lead the audience down the path of Barton's hallucination, Joel and Ethan Coen empowered Lievsay to create the sonic environment that is disconcerting and haunting. The scant use of Carter Burwell's music is tasteful and intriguing, but it takes the secondary role of both the segue and the sound bridge, rather than the more expected and traditional role of steering emotion and anticipation. Barton's spatial reality is created largely by the sounds he hears.

The story of *Barton Fink* revolves around the experience of a playwright and his excursion to Hollywood to "write for the pictures" after a successful run in New York theatre. Fink writes about the "everyman" and takes any opportunity to expound on his theories about the life of the ordinary man, even to his hotel neighbor, who embodies the very incarnation of the everyday workingman that Barton claims to admire. While Barton struggles to write his first script for the Hollywood studio that has brought him west, he discovers his neighbor may be the brutal killer "Mad Man Munt" and eventually suspects Munt may be after his own relatives back east. Barton encounters a variety of characters including a William Faulkner-like colleague, two detectives straight out of the "hardboiled detective" film noir world, and a quasi-love interest, Audrey, whose severed head may be in the package Munt left with Barton before he made his departure. Barton lives in one room at the hotel, and except for occasional visits to the studio to meet with the studio head and other ancillary characters, his only escape from his writer's room—and writer's block—is the picture of a bathing beauty at the beach that is posted above his desk.

The story that Ethan and Joel Coen wrote *Barton Fink* in three weeks while they were themselves experiencing writer's block during the development of *Miller's Crossing* (1990) may or may not be true. Foley artist Marko Costanzo says it absolutely is true, while his Foley mixer Bruce Pross says that is a modern myth. Regardless, the brothers did, at some point, decide to write about what they had experienced as difficulty in writing. It also seems to be true that they penned the screenplay quickly.[3]

Embracing an Alternative to Analog

The strangely subjective sound as it is evidenced in Barton's mind is in sharp contrast to the more pedestrian sound design Lievsay and the Coens employ in the exterior world of the more "real-world" characters. Whether Barton is alone literally or while among others, his inner world is expressed aurally and is masterfully executed by Lievsay and his colleagues. Lievsay's decision to design and edit the sound effects for the film digitally was a pioneering move for the sound designer. While Hollywood was grappling with the analog to digital transition with complexities involving union concerns, studio finances, and digital platform discrepancies, Lievsay had already been experimenting with alternatives to analog that would allow him more autonomy with his sound design and propel his desire to be more involved with the final mix.

Lievsay delved into digital editing gleefully. He had worked with the Synclavier, along with his colleague Eugene Gearty at Sound One, and was eager to use a stand-alone digital editing system. New England Digital Corporation, which had developed the Synclavier, introduced Post Pro—a hard drive editing system and Lievsay's editing tool on *Barton Fink*. Lievsay leapt into the new technology and availed himself to Post Pro for sound effects and Foley. However, dialogue editing was not possible on Post Pro, so Lievsay opted to have editor Phil Stockton continue with the standard analog editing practices for the production dialogue duties.[4]

New York postproduction sound crews tend to be smaller than either of those in Hollywood or the San Francisco Bay Area. It is not uncommon for sound editors to combine the tasks of editing sound effects, background effects, and dialogue, while a Hollywood sound crew most often separates tasks by specialty unless the film is specifically low budget. Additionally, in New York, it is more common for a film project to employ only one Foley artist, which is unique to New York work practices as compared to both the Bay Area and Hollywood—both of which tend to budget for two artists for the majority of major film projects.

Lievsay calls his approach to his profession the "dumb method," which for him means "do a good job and charge a reasonable amount."[5] He believes it was easier to get a foot in the door as a sound professional in New York this way. Prior to *Miller's Crossing*, Lievsay supervised all of his projects at Sound One, although he was not satisfied with the traditional practices on the mix stage: "It was a fairly clear subset of what was happening [in Hollywood]."[6] Lievsay prefers his own methodologies and had felt hindered by the standard practices of his colleagues. "I like working on the movie and making it as good as it can be and that's really easy to do on a smaller film. There are less people involved and less pre-conceived notions of what it should be like," reveals Lievsay.[7]

However, he was very happy to make use of the Synclavier at Sound One and was able to create interesting sound effects for his soundtracks. Lievsay, sound editor Eugene Gearty, and Foley artist Ron Bochar decided that building a Foley stage might finance their own Synclavier. What began as a "scheme" to fund his preferred sound design methods became a full-fledged sound editorial company that allowed Lievsay to employ other professionals (along with partners Bochar and Bruce Prosser) and offer major competition to Sound One. Before he had his own company, Lievsay explains how postproduction sound operated, and what prompted him to forge out on his own early in his entry into more creative sound design:

> In New York, people were taking library stuff and putting them on 6 or 7 tracks and mixers were taking whatever they preferred and when you went to the mix you handed your material to the mixer and from that point forward it was the

mixer's show. You would speak when spoken to and you had this and that. The style of sound editing was universal for a while and then it broke down. When a project was more abstract, the old ideas didn't cover it. You couldn't go to the library and get some sound design stuff and put it on tracks and work it out in the mix. You had to have an ability to have an idea and a concept and have these sounds conform to your idea. So, when you got to the mix and you were confronted with very conventional people . . . who were shouting you down, it was like a war. It's hard enough when you have an idea and you're being charged by the filmmakers to come up with something abstract to then go into a room with very conventional people and then try to convince them that your idea is a good idea and that it's worth it to invest the time . . . so, it's very clear that you need to be able to do that yourself.[8]

Creating an Independent Workspace

In 1989, they opened their new postproduction sound company, providing sound editing and Foley. From its inception, c5[9] began to edit film soundtracks digitally.[10] "We'll be the only studio on the East Coast, excepting Sound One, to have the capacity to mix and record an entire film on computers,"[11] Lievsay commented in an interview for Village Voice in 1989. "We'll still be doing our final mixes at Sound One, and since that's where the biggest profit margin is, they shouldn't be too upset."[12] Before long, it became possible to do some mixing within the same digital system. For Lievsay, digital editing was an expansive experience: "It allowed me to be liberated from old ideas."[13] To clarify, Sound One had the Synclavier for digital sound design, but Lievsay was projecting into the future when his company and Sound One would be mixing digitally as well. It was in actuality a few years later before final mixes were executed with digital technology. Lievsay began premixing his edited tracks in his own computers as soon as the capability was available.[14] He explains,

> It wasn't until we started working with Synclavier that we were able to have a sophisticated combining of elements and actually mix sounds together. We were restricted to the physical limits of the machine. We were getting to the point where we had to premix reels repeatedly just to combine all of the elements into a manageable form. Now we can do anything we want in combining and mixing sounds. You have tremendous physical control over the sounds.[15]

With analog editing, all premixes were performed on the dub stage by the rerecording mixer, which Lievsay found daunting:

If the mixer happens to raise a fader incorrectly on something you may have intended to be a subtle element, which might work very well, the director goes, "What's that supposed to be? We don't need that," and a valuable sound effect that you really thought contributed is now out of the picture. It really would never be given a chance under the old system. It can take more momentum away than you would ever possibly imagine.[16]

Thus, Lievsay is clear that he prefers to pre-mix his sound effects:

Miller's Crossing was the first movie where we were able to compress all of the sound effects and backgrounds down to eight or ten stereo pairs, and we've been doing that ever since. That's one contribution the [electronic editing] machines have—you can do a lot of that compressing yourself. You can establish the balances yourself. You're developing a scenario. This way you can bring something which has a consistent point of view and it's your opinion. Your opinion of what the scenario should be is as valid as everyone else's opinion.[17]

What Lievsay is expressing is a critical cultural difference between New York and both Hollywood and the Bay Area postproduction sound communities. While the West Coast film enjoy a more collaborative environment between the rerecording mixers and the supervising sound editor or sound designer, New York, true to Vorisek's tutelage, had developed a hierarchy of a mixer-controlled environment that was amenable to the rerecording mixer—and allowed the mixers to become influential with the directors—but crushed the creative juices of an artistic designer like Lievsay. His need to create and control his art propelled him and his colleague, Gearty, toward the more flexible and editor-friendly electronic tools.

The main task of the rerecording mixer, however, was left to Lee Dichter at Sound One. Lievsay adds, "Lee [Dichter] did a very good job."[18] Thus, while the sound effects recording, editing, and premixing was executed digitally, the final mix was completed in analog—as was still the only available option in 1991.

Dichter had been a veteran New York film industry professional from his early years working with his father in commercials. He learned to mix dialogue first in commercials, then in documentaries. Six of the documentaries won Academy Awards. Dichter's affinity for mixing dialogue derives from his days massaging the dialogue in both genres.[19]

Atypical Collaboration

When beginning his work on *Barton Fink*, Lievsay did not have the typical spotting sessions with the Coens. Instead, he and the Coens went over the material—what kind of ideas they had—then Lievsay "went into my cave."[20] They communicated regularly about Lievsay's presentations of scenes, and Lievsay would implement the Coens' comments or notes. Sometimes there would be specific notes on Foley or sound effects. It was a process of building—which suits Lievsay, as he had intended in his youth to be an architect. "I could make the stuff and play it back in a mixed condition," which for Lievsay was real freedom. "It was clear I was presenting my very specific idea of what I thought it should be. Then we could get into some very specific terms of how to adjust it." Lievsay then adds, "They [the Coens] had their opportunity to vet the material and then we just needed Lee [Dichter] to put it in the movie. That was a real advance in our collaboration."[21] The following quote from an article in volume 1, issue 1 of *Soundtrack*, in November of 2007, illustrates the very special relationship the Coens have established with both Lievsay and composer Carter Burwell:

> The Coen brothers' approach begins at the script-level, in that they fill their screenplays with descriptive sound scenarios and phonetic dialogue. It is at this point that the Coens invite Skip Lievsay, their regular supervising sound editor and mixer, and Carter Burwell, their regular composer, to review and comment on the script. These pre-production discussions are often minimal. They usually focus on the practical needs of the film and the overall themes of the narrative. Meeting early in the production allows Lievsay and Burwell more time to consider the sound ingredients they are to contribute to the film. What is more, as the Coens rarely veer from their original script during filming, they can be confident that a large majority of their initial ideas will not be wasted. After sketching out ideas generally, Lievsay and Burwell meet with the Coen brothers to spot the film; this is usually done at the rough cut. During these sessions, sound effects and music are explored simultaneously, reducing the risk of conflict in the final mix. Following this, Lievsay and Burwell continue at some level to exchange ideas that will complement the needs of the film. When the elements are eventually delivered and the tracks are laid in, the result is a soundtrack that not only fully complements the narrative, but it is also integral to the film's construction. As a result of this approach, Joel and Ethan Coen have formed a close bond with their sound personnel. Lievsay and Burwell

have been asked to work on all of the Coen brothers' films, regardless of the greater communication between all those responsible for the film. It encourages a greater familiarity in working habits and a higher level of trust. Above all, it suggests artistry over ego in that it emphasizes the need to focus on the end product rather than one's individual contribution.[22]

Lievsay believes strongly that transitioning to digital editing has enhanced his process greatly:

> It's a real team effort. Everyone really does work together, but unfortunately, mechanically, it's a very big process. It takes a long time. We feel charged by the directors and mixers to do anything we can do to make it go as efficiently, cheaply, and cleanly as possible. These machines have really contributed greatly to that end ... Because of the amount of material you can play at the same time, you get a very, very good idea of what the overall final result could be. Also, you get to weed out the sounds that aren't working, and you get less attached to what gets taken out. You become more focused on the important critical elements—everyone wins.[23]

Stepping Up to the Foley Plate

Bruce Pross was the Foley mixer on the new Post Pro system for *Barton Fink*. Previously an assistant editor and recent sound editor, he was trained on Post Pro for the purpose of mixing the Foley for artist Marko Costanzo. "I had never engineered anything in my life," says Pross. "I didn't have much to compare it to. It was really frustrating, as I recall. The thing kept crashing. It looked like the monolith from *2001*. It was a tower, 3' by 3' by 7' tall!"[24] Pross remembers that the Foley sessions had to be stopped regularly to allow time for the Post Pro to cool. He would keep a thermometer behind him and glance back to check the temperature. When the mixing booth got too hot for the Post Pro, Pross would stop recording, turn on the air conditioner for a few minutes.[25] Post Pro worked like a 24-track recorder[26] in that it could record "samples" and put the audio where it belonged. While Pross wonders if perhaps c5 might have had an easier time recording Foley on a 24-track analog machine—which was the standard at that time—according to Costanzo, Pross was able to bring the cues up faster and manage the recording of cues quickly and efficiently on the Post Pro system.[27]

The Post Pro 16-track recorder for the Foley stage, one 4-track Post Pro editing machine, and a used Synclavier cost c5 an initial outlay of $148,000. Pross still has the invoice from this first purchase for the delivery on August 23, 1989. He remembers that he edited some of the Foley on *Barton Fink* at night to help defray some of the cost of the machines. The stage recorder had four separate 4-track hard disks that would insert into the machine. Each one could be replaced with another 4-track disk. He would hand deliver the hard disks to an editor who would then edit on the 4-track Post Pro machine.[28]

The Foley stage was of particular concern to Bochar and Lievsay. Lievsay was involved with the design of the Foley stage, which was more "live"[29] than the room at Sound One.[30] Costanzo refers to the room where *Barton Fink* Foley was mixed a "Halo of Sound,"[31] as he prefers the liveness of the room.[32] Since most of the film takes place in Fink's hotel room, the liveness, to Costanzo, is a particular asset. What Pross remembers about the Foley stage is that a rock and roll band rehearsed in the room above them and that they had to record around the rehearsal times.[33]

Costanzo remembers working on the Foley for *Barton Fink* for about six to eight weeks. As he recalls, this was a normal schedule for major films back in that era of filmmaking. It should be noted that a comparable film in Hollywood might have warranted three to four weeks of Foley stage attention in 1991. The Coens were very involved with the Foley. "The Coens wrote all the sound cues," recalls Costanzo.[34] Few films have the attention of the director(s) regarding sound design that any Coen Brothers films garner. Eugene Gearty confirms that the Coens are quite specific with their sonic descriptions. "It was scripted ... and Joel and Ethan did that," says Gearty. While the imagination, execution, and overall cohesiveness of the sound design of *Barton Fink* stems from the creativity of Lievsay, Costanzo, and the other sound professionals, the clarity the Coens brought to the script was critical to the inception of the sound design.

Sound Evokes the Uncanny

Barton's hotel room is the setting for much of the narrative. His temporary residence is the Hotel Earle, a dilapidated residence that Barton has rented while he is in Hollywood, writing a script for a wrestling picture for a major studio. As Barton checks into his new "digs," and chooses to live apart from the more elite domiciles of his colleagues, the now infamous bell ring is introduced (Plate 2). Lievsay used a sound effect that was only fifteen seconds long, but the Coens wanted the bell to last much longer. After several complicated attempts to work with elaborate sounds and reverb to accomplish this signature effect, Lievsay was not getting what was required. "It was metal with non-harmonic overtones, and I needed to get it down to its basic components."[35] Lievsay credits the digital hard drive system for making this sound design challenge possible. Utilizing both the Synclavier and the Post Pro systems, Lievsay was able to create, manipulate, and edit sounds more seamlessly than previously when he would design sounds on the Synclavier, then transfer the sounds to magnetic film for editing.

Costanzo laughs about the Foley involved with the three major wallpaper cues in the film. Remembering that the Coens were enthusiastic about the importance of the wallpaper, he remarks, "Viscous was the word of the day!"[36] Lievsay remarks that the Coen's wanted the wallpaper to "have a life of its own."[37]

The well-known signature sound, Barton's hotel room door, brings a chuckle to Lievsay as he relates the story of its accidental design (Plate 4). The tight suction sound that sucks Barton's door closed was, in actuality, the door to the Foley stage that kept outside noises from bleeding into the stage. Lievsay had put the door in as a joke because in his words, "Both Ethan and Joel laugh like donkeys."[38] When the Coens heard the sound, they responded with the laugh and asked, "What's that supposed to be?"[39] In the next scene, when John Goodman's character visits Barton in his room, the door sound was missing. The Coens wondered where the door sound was.[40] They wanted to keep it. "We used it so many times—that door was opened and closed a hundred times in that movie."[41] Lievsay's gag became arguably the most discussed signature sound in the film. Its importance—sealing Barton into his room and keeping all outside influences out—becomes a story point and contributes to the isolation and containment of Barton's room.

Pross states, "I've always loved working on their [the Coens'] movies." He remarks on their ideas regarding specific sound ideas in Barton Fink: "They wrote the thing. The scene calls for it; demands it. But they're open enough to get your juices flowing."[42] He remembers recording Costanzo's stomach growls as a sweetener for a scene in Ben Geisler's office. The character is "all agitated" and "Marko's stomach gurgles were great for his [Geisler's] stomach." Another sweetener Pross recalls came out of a surprise visit from Joel Coen who was visiting Lievsay across the hall. Coen wanted an additional sound in the hotel elevator that was a rubbing or hitting against the inside of the elevator (Plate 3). Pross calls this encounter one of those "memorable times when I pleased Joel and Ethan"[43] with what they designed on the Foley stage to satisfy that augmentation to the quality of the elevator. Pross also confirms the signature Coen "hubcap" sound: the settling of a prop that mimics the motion of a fallen hubcap as a comic ending to a moment in a scene.[44] One example of this "hubcap" is when the character of Lipnick (Michael Lerner), the studio head, walks into his office. "At one point he slams his hand down on the desk, and you hear what's essentially a glass ashtray—that's a hubcap" (Plate 7). Lievsay continues, "My Foley guys put hubcaps in everywhere . . . It's a little joke."[45]

Musical Chairs

Composer Burwell and Lievsay collaborate on all Coen brothers projects. Regarding *Barton Fink*, Lievsay says, "We work together. We did not have to fight for sonic space."[46] Dichter believes that Lievsay's sound design "was almost a [film] score in some places."[47] "The sound effects were integrated with the music. We work together on these issues," adds Lievsay about his relationship with Burwell. This relationship is typical according to Burwell:

> A film mix is a battlefield, generally—dialogue people fight with the sound-effects people, who fight with the music people. In the end, nobody's completely happy, and if you've been to those mixes, you can sometimes anticipate what's going to happen and try to make sure that the music is not too badly bruised. But it's a battle."[48]

Lievsay shares Burwell's skepticism in general, "It's a dangerous territory," but adds, that with Burwell, "it's a real collaboration."[49] Burwell also comments on his philosophy about film music, and how it fits into a narrative, "I think the function of music is to show you something you don't see onscreen."[50] Burwell's use of music in *Barton Fink* is minimal and his orchestration is scant, which adds to the sense of uncanny that permeates the character's hotel room and Fink's questionable sanity.

The unique relationship of Lievsay, Burwell, and the Coen brothers as an example of collaboration of artists illustrates what is possible in the New York film community. Rather than impose artificial categories of above-the-line and below-the-line, as is the practice in Hollywood, the goal for these artists is filmmaking, and the lack of hierarchy makes the practice of collective authorship bear fruit. However, the very nature of the close relationship between sound professionals and directors in New York is historically traditional. Directors work with the same picture and sound editorial professionals from project to project. Unlike Hollywood, films are not typically package deals constructed by and overseen by larger corporate entities. The artists themselves tend to determine the timelines and outcomes of the film projects.[51] The New York film community is unlike the Bay Area in that it is fast, economical, and works primarily with New York filmmakers. New York is a film community that has developed "a practice of working for a particular artist, unlike Hollywood, which is an industry."[52]

Getting Noticed

Until *Barton Fink*, a New York soundtrack, supervised by a New York sound editor, had not gained significant attention of the mainstream Hollywood sound community. In 1992, as the Motion Picture Sound Editors (MPSE) were selecting their nominations for best sound editing, three of Lievsay's films were considered: *The Silence of the Lambs*, *Goodfellas*, and *Barton Fink*. Ethel Crutcher,[53] the organizer for the program, asked Lievsay which one he would prefer to be the nominee since he could not have three. "I chose *Barton Fink* because I figured I would rather lose with that one than one of the other two."[54] He won the Golden Reel that year.[55] The other two films were edited on analog and mixed at Sound One. Lievsay's work on *Barton Fink* was also the only one of the three edited on Post Pro, and which had allowed Lievsay to display a more innovative soundscape.

Skip Lievsay and Eugene Gearty can be credited in large part for the faster transition into digital editing in the New York postproduction sound community. Prior to Lievsay's entry onto the scene with his three groundbreaking soundtracks in 1991, the New York sound community was not taken seriously by any Hollywood elites. Most of the established sound professionals in New York had worked on "dialogue films"[56] and not films known for sound design elements. After *Barton Fink*, everyone in the American film sound community knew who Skip Lievsay was, and was aware that New York feature films had moved into electronic sound editing. After the editors at c5 began editing digitally and "mixing in the box,"[57] Hollywood and the Bay Area—both adopting feature film digital sound editing after Lievsay—knew it was only a matter of time before they would have to learn how to do the same. As *The Conversation* had changed the conversation regarding sound design, so had *Barton Fink*.

Gearty, who was responsible for introducing Lievsay to the Synclavier originally, but who did not work on *Barton Fink*, explains a key reason he and Lievsay were so excited to get into electronic editing as soon as possible, "You can have a machine that's quick at doing what you want but now you have twice as much time to be creative with it."[58] This is a key distinction from Hollywood, where the efficiency of the digital audio workstations (DAWs) was seen as an opportunity to use fewer workers for the same amount of time. He continues, "You can bring bodies to the table and it works [referring to the Hollywood system of specializing editing tasks and utilizing large crews]. But films were handcrafted in New York and Chicago and the Bay Area."[59] Gearty believes that the reason he and Lievsay were able to adapt to the digital editing systems more quickly was because the financial commitment would have been too large for any large facility that employs large crews. "The stakes are high. It takes a lot of money." He also explains why New York editors began premixing within the editing machines earlier, "Guys like me and guys in the Bay Area were mixing and editing simultaneously. And the structure of the big mixers in L.A. would never allow that."[60]

Lievsay expresses the crucial relationship between the sound designer and the director in New York:

> It has a lot to do with our relationship with Joel and other directors in that we're able to see what they have in mind and then come up with an effect like that. The directors have to be able to be magnanimous enough to share what they're doing and not simply recruit the specific things they had in mind.[61]

"I'll Show You the Life of the Mind!"

Barton's hotel room is the setting for much of the narrative in the film. Subsequently, most of the sound effects discussed above reflect the goings-on in his room, and thus, this textual analysis examines the hotel as his domicile, utilizing Lievsay's sound design and Costanzo's Foley, which together create a subjectivity to Barton's room that reflects his writer's mind, his perceptions of events, and his relationships with others.

His temporary residence is the Hotel Earle, a dilapidated residence that Barton has rented while he is in Hollywood, writing a script for a wrestling picture for a major studio. Barton is a New York intellectual playwright, out of his element in this new city writing for a new genre. While he sits at his desk, facing the almost blank page, occasionally allowing himself to mentally leave the room by looking up at a photo of a young woman sitting at the beach, he hears a disconcerting and percussive sound that alerts us and brings him back into his depressing environment. It is the wallpaper over the head of his bed, beginning to peel away from the wall.

It should be noted that the Coens incorporated the wallpaper as an additional character in the film with deliberation. In an ordinary film, what covers the wall would be a matter of art design, but only in stories like "The Yellow Wallpaper,"[62] or films such as *Repulsion* (1965) or *The Conversation*, do walls and their coverings matter as much as in *Barton Fink*. In fact, if one watches the film closely, the opening scene in the New York theatre where the protagonist is first introduced mouthing his own words as the actor recites them—and the actor playing the character sounds suspiciously like Barton himself—it is clear that the wallpaper of the theatrical set resembles the wallpaper at the Hotel Earle. The importance of the wallpaper causes the viewer to wonder, why does the actor in the play sound like Barton and dress like Barton as we see him later in the film at the hotel? Is this film out of sequence as a further indication of Barton's insanity? Or, as is more likely, Barton has dressed as his lead "everyman" character in his play, and deliberately booked himself into a hotel that represents where "the common man" would live, so he can avoid cutting himself off and connect more with his self-proclaimed hero.

It is clear that the room is hot, because Barton has a table fan blowing. However, the wallpaper has a "sweat" all its own. The sound, which alerts Barton to the event, is sharp and stiff, accompanied by a "goopy glop" that is disturbing. Barton turns to watch the slimy paper, which is seemingly alive and menacing, abandon its function of cover and protection. He watches for a full twenty seconds before he gets out of his rickety and creaky chair, stands on his squeaking and sinking bed, and proceeds to confront the intrusion. With his left hand, he presses the thick paper back onto the wall, the audible ooze continuing to seep out (Plate 5). Barton examines the mucous texture on his hand and decides to smell it. For the first time Barton finally really looks at his surroundings: his "home" for the first time. He becomes aware of the unsettling ambience of his self-imposed confinement.

In cinema, sound can evoke our response to an image before we see it. The aural representation can portend an uncomfortable moment. The eerie quality of Barton's wallpaper is created acousmatically[63] as he hears the belching goo before he sees it. Barton first hears the intrusion, aware of its misplaced role in his writer's mind, and only after he turns in his chair, does he visually register what event is taking place.[64] "The acousmatic sound maintains suspense, constituting a dramatic technique in itself."[65] Barton's writing as well as his stability are distracted aurally at first. He is aware of the spooky sound before he can identify it as the wall's epidermis. The ooze and unclothing would scarcely warrant attention had he not been first alerted to the ghostly sonic aura. The suddenness of the first detachment, along with the ectoplasmic secretions, establishes the sinister intentions. No longer is he going to rest comfortably in this room. Barton's vigilance increases, both visually and auditorially.

Kevin L. Ferguson, in his article "On Wallpaper in Some Films,"[66] explores how wallpaper evokes further engagement in film. "Each time a character leaves one room brings a new question; not 'where are they going' but 'what kind of room will they see when they get there'? And what will I see and what pleasure do I receive from these variegated rooms?" The wallpaper in Barton's room demands attention. Rather than play its particular role as part of the set design in the mise-en-scène, it is a supporting player in the narrative itself. Not only do the characters specifically discuss the significance of the wallpaper three times, but its looming uncanniness is omnipresent throughout the film (Plate 6). "Wallpaper in the cinema certainly must be associated with the neorealist tradition, with the use of deep-focus photography and the concurrent invitation to look farther into the background. The invitation to look—at what you will when you will—comes at a cost, for suddenly each component of the frame becomes important."[67]

Barton's hotel digs hail a time past, when details like wallpaper and wooden banisters were the norm in fine hotels. The Hotel Earle, while decrepit in Barton's story, contains a sense of history and nostalgia. It stores secrets that are revealed with every inch of the wallpaper's disrobing. Finnish architect Juhani Pallasmaa explores the metaphysical meanings associated with inside and outside spaces by comparing the connections between our own bodies and buildings. He argues that we are in "constant interaction with the environment; the world and the self-inform and redefine each other constantly . . . there is no body separate from its domicile in space, and there is no space unrelated to the unconscious image of the perceiving self."[68]

Barton's wallpaper is aged and disintegrating, yet it protects the walls from complete nakedness. The protection, much like skin, betrays its purpose as it pulls away from its surface, with its congealing viscosity. "This fear of the traces of wear and age is related to our fear of death."[69] Perhaps as the skin of the walls disintegrates, Barton feels a shudder as he recognizes that he, too, will crack, peel, and dissolve into a powder—dust to dust.

As Barton attempts to reconnect his wall coverings to the wall he shares with his lovemaking neighbors—using tacks sent up by "Chet, my name is Chet," he once again is interrupted by Charlie. Now Charlie shares that he has an "ear infection . . . chronic thing. I have to put cotton in it to stop the flow of pus." No sooner has Barton closed off one aspect of his interior life that is leaking that he encounters another more insidious reminder of his own vulnerability. "Can't trade my head in for a new one," Charlie muses. "Yeah, guess you're stuck with the one you've got," replies Barton. Every moment in Barton's room is painfully empty with each sound surrounded with emptiness and blank space like the blank page on his typewriter.

Gaston Bachelard (1994) discusses the importance of our shelter, our "hut":

> All really inhabited space bears the essence of the notion of home ... the imagination functions in this direction whenever the human being has found the slightest shelter: we shall see the imagination build "walls" of impalpable shadows, comfort itself with the illusion of protection—or, just the contrary, tremble behind thick walls, mistrust the staunchest ramparts.[70]

In his temporary writer's retreat, Barton expects a feeling of home, solitude, sanctuary. The walls he inhabits are his refuge, even if only temporarily, from the society that invades his "life of the mind." "Every corner in a house, every angle in a room, every inch of secluded space in which we like to hide, or withdraw into ourselves, is a symbol of solitude for the imagination; that is to say, it is the germ of a room, or of a house."[71] Barton's desire to write his script is disrupted continually by noisy neighbors, annoying mosquitoes, and his room that methodically conspires to preclude the meditative environment so necessary for Barton's imaginative inventions. It is the aural confinement he has created that imprisons him as much as the narrow minds of the studio executives or the circumstances of his questionable neighbor.

Barton's world becomes more surreal as he is implicated in a murder that occurs in his room. The audience becomes fairly certain the murder is the action of his next-door neighbor. At this point in the film, hyperbolic events seem almost expected after viewers are conditioned to accept the bizarre as predictable in Barton's hotel "cell." Lievsay's sound design ensures that audiences are never really certain whether the room's sounds are part of the objective diegesis or rather the subjective experience of Barton's imploding mind. However, they are real to Barton and contribute to his utter inability to retain a sane focus on the task at hand.

At various times in the film, when Barton can no longer tolerate his restrictive cloister, he "escapes" by staring at the photo of the young woman, sitting on the beach. The accompanying soundscape is one of the smooth crash of ocean waves and seagulls conversing with one another. Again, it is ambiguous whether his desire is for the fantasy woman or for freedom from his writer's prison. Barton, however, allows himself momentary relief with an imagined sense of freedom from his confinement.

Barton self-isolates. He decidedly finds a small space that he believes will not distract him from his writing. His first encounter with the peeling paper, expelling its mucilaginous afterbirth, commences our own surrealistic journey with Barton into a world of slime, exposure, sexual inhibitions, violation, penetration, intrusion, and the uncovering of closeted skeletons. While Barton sequesters himself, he does not try to avoid the aural accosting that he experiences and the unsettling effect on his psyche. The writer's curiosity and fascination with the imaginative seduces him.

Near the end of the film, as Barton returns to his room to discover the two detectives reading his script, he passes by the wallpaper now peeling totally off his walls. His room is disintegrating. "Why is it so goddamn hot out here?" asks one of the detectives? They see fire flaming up from the elevator. Mad Man Munt steps out of the elevator and begins his rampage, "Look upon me. I'll show you the life of the mind!" After he kills the two detectives, Charlie (Munt) visits Barton one last time and says, "You're just a tourist with a typewriter. I live here. Don't you understand that?" Charlie breaks the bedframe where Barton has been handcuffed by the detectives and finally frees him. Charlie returns to his room while the hotel continues to burn. Barton, however, walks out of his room, and leaves the Hotel Earle.

At the end of the film, at the ocean's edge, Barton has survived the heat in his room, the assaulting noises surrounding him, the concealing of a dead woman's body, his neighbor's reappearance as "Mad Man Munt," the constricting walls, the wallpaper's striptease, and Dante's Inferno, and has broken through his writer's block. Yet still, he faces the rejection of his triumphant script. He experiences another confinement—his true straitjacket: Barton's own labyrinth created by his "life of the mind." Near the end of the film, the pathos of Barton's words is delivered to the studio head: "I tried to show you something beautiful; something about all of us."

Lievsay, Costanzo, and Pross—utilizing the descriptive scripting of the Coens, and with the cooperative collaboration of Burwell—constructed an aural interiority for *Barton Fink* that resonates with the self-absorbed writer who leaves the audience with those words of pathos. The increased control over the sound design that Lievsay and his team acquired by incorporating the digital editing systems into the postproduction sound editing workflow allowed them more fluidity with the process and encouraged Lievsay to fulfill his desire to create a bolder and more imaginative sound design. Rather than continue the tradition of the rerecording mixer determining the trajectory of the soundscape, Lievsay altered the convention of sound design in New York by presenting his entire sound design fully rendered prior to the mix. He left analog sound editing behind and pushed into the future of digital sound editing for feature films before either Hollywood or the San Francisco Bay Area made the transition.

Notes

1. Lee Dichter Personal Interview, interview by Vanessa Theme Ament, Phone, in New York, June 17, 2014.
2. Joel Coen and Ethan Coen collaborate. Joel is the director of the team and always employs Lievsay as his sound designer.
3. Both Costanzo and Pross are certain of their facts. In my interviews with them, I have found they have contrary memories of many occurrences on the Foley stage as well. Costanzo remembers the experience of doing the Foley for *Barton Fink* as easy and fun, while Pross remembers the technical issues and the time-consuming aspect of conforming to the new editing technology. Both, however, remember working with the Coens and working on any Coen project as wonderful and fulfilling. Marko Costanzo, Personal Interview, interviewed by Vanessa Theme Ament, Skype, in Fort Lee, NJ, June 23, 2013; and Bruce Pross, Personal Interview, interviewed by Vanessa Theme Ament, Skype, in New York, June 10, 2014.
4. Eugene Gearty, Personal Interview, interviewed by Vanessa Theme Ament by Phone, August 18, 2013.
5. Ibid.
6. Ibid.
7. Ibid.
8. Lievsay, Personal Interview (2013).
9. The name of the company comes from "Chelsea Five." The building was in Chelsea and there were five original owners. Bruce Pross had to shorten it for the

incorporation papers, thus, c5. This is how Bruce Pross relates the story. Bruce Pross, Personal Interview.
10 Lievsay had begun to work with the Synclavier at Sound One but felt limited by the mixers that were "old-school," as he calls it.
11 Amy Taubin, "A Sound Is Built," *Village Voice* 34 (October 3, 1989): 71–2.
12 Ibid.
13 Skip Lievsay, Personal Interview (2013).
14 Ibid.
15 Vincent LoBrutto, *Sound-On-Film*: *Interviews with Creators of Film Sound* (Westport, CT: Praeger, 1994), 266
16 Ibid., 267.
17 Ibid.
18 Skip Lievsay, Personal Interview (2013).
19 Lee Dichter, Personal Interview.
20 Skip Lievsay, Personal Interview (2013).
21 Ibid.
22 Randall Barnes, "The Sound of Coen Comedy: Music, Dialogue and Sound Effects in *Raising Arizona*," *Soundtrack* 1, no. 1 (March 2008): 15–16.
23 LoBrutto, *Sound-On-Film*.
24 Bruce Pross, Personal Interview.
25 Ibid.
26 The 24-track was a tape recorder method that had multiple tracks to record analog on Foley stages and rerecording stages at this point in postproduction.
27 Marko Costanzo, Personal Interview.
28 Bruce Pross, Personal Interview.
29 Foley stages are designed to allow sound waves to move. Most include soft "traps" on the walls to deaden the sound. The Foley stage at c5 allows more life to the sound, which is a more traditional practice in music recording studios. The result is that sounds are recorded with more interior sounding qualities. To alleviate this if desired, the Foley artist must use baffles—foam structures—to contain the sound of some recorded cues. For further information on Foley stages, please refer to *The Foley Grail: The Art of Performing Sound for Film, Games, and Animation*, 2nd ed. (Burlington, MA: Taylor & Francis, 2014), 181–206.
30 Marko Costanzo, Personal Interview.
31 This is Costanzo's term.
32 Marko Costanzo, Personal Interview.
33 Bruce Pross, Personal Interview.
34 Marko Costanzo, Personal Interview.
35 Skip Lievsay, Personal Interview (2013).

36 Marko Costanzo, Personal Interview.
37 Skip Lievsay, Personal Interview (2014).
38 Ibid. (2013).
39 It is commonplace in Hollywood for sound editors to put joke tracks in for directors with whom they are comfortable. This story of Lievsay's illustrates the same is true for him and the Coens.
40 Skip Lievsay, Personal Interview (2013).
41 LoBrutto, *Sound-On-Film*, 257.
42 Bruce Pross, Personal Interview.
43 Ibid.
44 This is the Coen version of a "comic button" to a prop sound.
45 LoBrutto, *Sound-On-Film*, 1994.
46 Skip Lievsay, Personal Interview (2014).
47 Lee Dichter, Personal Interview.
48 Tod Lippy, "Carter Burwell," *Projections*, 2000, 51
49 Skip Lievsay, Personal Interview (2014).
50 Lippy, "Carter Burwell," 46.
51 Tod Lippy, *Projections 11: New York Fill-Makers on New York Film-Making* (London: Faber & Faber, 2000).
52 This is how Lievsay characterizes the difference. Dichter and mixer Tom Fleischman agreed.
53 Ethel Crutcher was an active part of MPSE. While not officially a member, she had been married to Norval Crutcher, a respected sound editor and former president. Mrs. Crutcher organized the MPSE awards banquets for many years and took on many responsibilities as part of her duties. She asked Lievsay this question because he was considered for three major films and this was not normal for the event.
54 Lievsay remarked during the interview on August 18, 2013, that he did not think anyone would appreciate his work on *Barton Fink* because it was so strange and not the "big and loud" kind of soundtrack that usually wins sound awards. Thus, he did not go to the ceremony. Dody Dorn, a fellow editor, called him to tell him he won during the evening after his win was announced. Skip Lievsay, Personal Interview (2013).
55 Source: Personal MPSE Program.
56 This is an industry term that refers to narrative films that rely on story or character rather than action, fantasy, horror, or music as a major component.
57 "Mixing in the box" enabled editors to do some minor premixing while they edited and before the final mix. This practice occurred more frequently after New York and Hollywood merged unions into one for both editors and studio mixers. This

practice is still not completely accepted by either all sound editors or mixers. But economics often requires mixers to edit and editors to mix.
58 Eugene Gearty, Personal Interview, interview by Vanessa Theme Ament, Skype, in Beaufort, SC, May 22, 2014.
59 Ibid.
60 Ibid.
61 LoBrutto, *Sound-On-Film*, 257.
62 Charlotte Perkins Gilman, *The Yellow Wallpaper and Other Stories*, unabridged ed. (Mineola, NY: Dover, 1997).
63 Michel Chion, *Audio-Vision: Sound on Screen*, trans. Claudia Gorbman (New York: Columbia University Press, 1994), 72.
64 Christian Metz's discussion in "Aural Objects" supports the notion that when we talk about film sound, "we are actually thinking of the visual image of the sound's source." Elisabeth Weis and John Belton, eds., *Film Sound: Theory and Practice* (New York: Columbia University Press, 1985), 158.
65 Chion, *Audio-Vision*, 72.
66 Kevin L. Ferguson, "On Wallpaper in Some Films," *Bright Lights Film Journal*, accessed June 17, 2014, http://brightlightsfilm.com/.
67 Ibid., 12.
68 Juhani Pallasmaa, *The Eyes of the Skin: Architecture and the Senses*, 2nd ed. (City: Academy Press, 2005), 40.
69 Ibid., 32.
70 Gaston Bachelard, *The Poetics of Space*, 1st ed. (Boston, MA: Beacon Press, 1994), 5.
71 Ibid., 136.

4

"How Would You Like to Work on a Monster Movie?": The Expressionism of *Bram Stoker's Dracula*

Sound that was both organic, yet magical.[1]
—David E. Stone, supervising sound editor

The fascinating story behind Francis Ford Coppola's adaptation of *Bram Stoker's Dracula* (1992) has been presented in various publications—as well as the special features on the DVD. This director had a vision that demanded fidelity to the spirit of the novel, while incorporating visuals that worked on a metaphysical level with the audience. Rather than embody a singular vision of an individual, with every element carefully controlled so as to fulfill a predetermined direction, the production was a confluence of "gathering the mushrooms for this pie."[2] The process involved collaboration on every level, with each step in production contributing invention and artistry neither necessarily anticipated nor previously attempted in a film of such extravagance. *Bram Stoker's Dracula* (*BSD*) was a culmination of a collaborative sequence of collective authorship that yields the bountifulness of artistic expression witnessed in the final product.

The narrative journey involving the film's soundtrack was a compelling one. *BSD* was produced during a time of technological convergence, which created problems from beginning to end. How the sound design was imagined, crafted, and executed for the film is consequential because it illuminates the chasm between those who were ultimately more powerful at Columbia and Sony Pictures, and the postproduction sound professionals who rarely—if ever—came in contact with the individuals who decided their salaries, working conditions, and, more importantly in this case, their working tools. In contrast to the previous chapter as the initial technological disruption was instigated by the creative sound designers in New York themselves, for reasons of independence and artistic expansion, *BSD* is a story of regime change, mistrust, and political economy, that drove a studio to insist the creative sound professionals work within constraints that continually vexed young aspiring talents and veterans alike.

Because Hollywood was grappling with the transition from analog sound to digital in the late 1980s and early 1990s, Coppola, Sony, Columbia Pictures, and sound professionals produced the film *BSD* with full (if apprehensive) knowledge that the industry was already beginning this technological transition. Factors including the politics between Columbia Pictures and Coppola, Sony's economic power over film professionals, and the collaborative culture of sound professionals combine to illustrate the overwhelming task of creating a coherent and artistic sound design—a task that implicated a plethora of obstacles resulting from the confusion and difficulties of navigating too many conflicting technologies that had to be wrestled together. Veteran professionals with little or no experience with the emerging technologies, and tech-savvy neophyte editors who lacked significant filmmaking experience, were expected to collaborate and develop a soundtrack worthy of Coppola's directorial vision. Working against a shortened schedule, an underbudgeted and nearly unprecedented sound design challenge, and a pioneering collaborative approach between Los Angeles and Bay Area sound professionals under the direction of a brilliant director who changed his editing choices up to the eleventh hour, the sound team managed nevertheless to create a film that won the Academy Award for Best Sound Effects Editing.

Yet Another "Dracula" Adaptation

Francis Ford Coppola's adaptation of *BSD* is an unusual one in several ways. The script went through several permutations, but screenwriter James V. Hart was determined to develop a script that was true to the novel and included characters ignored in previous film incarnations. The typical response Hart would get when he "shopped" his script was "why another Dracula?" Hart was convinced his script was true to the spirit of Stoker's novel. He also added historical information about the real Dracula, Vlad the Impaler, which was intended to add sympathy and compassion to the vampire's story. Rather than create a monster with no purpose other than to be a predator in survival mode, Hart hoped to bring to the screen a tragic and broken hero who had lost his soul. His choice to focus on the romance between Dracula and Mina captured, for Hart, the true spirit of Dracula—one that had not been told cinematically.[3]

Winona Ryder had previously been cast in Coppola's *The Godfather Part III* (1990) and had dropped out of the production for health reasons. Ryder had read Hart's script and asked Coppola to read it. She was interested in playing Mina and wanted his opinion of the script. Coppola decided he wanted to direct it.[4]

Coppola's adaptation of *BSD* was intended to exhibit more fidelity than previous incarnations. Close fidelity is typically not the primary goal of adapting a book to a film. However, Coppola was a fan of the novel, and had always been frustrated by the lack of attention to many of the details in Stoker's rich text. He and Hart agreed that the journal and letter writings of the novel were crucial and needed to be included.[5] While Hart did fulfill much of Coppola's vision with a script that incorporated characters and side stories that other films had ignored, the original screenplay provided at least a basis for how Coppola wanted to transform this classic novel into a more historical narrative that could be recognized as the quintessential Dracula myth in film.[6]

Coppola's unusual approach to directing is well known within the industry.[7] He rehearses scenes tirelessly before shooting, encourages odd improvisational exercises with his casts to free up characterization and relationships, and works closely with all artisans and craftspersons involved with the production.[8] He often includes family members in his enterprises and encourages "working without a net"[9] for the sake of creative interpretation. What is unique about Coppola's direction of *BSD* is his approach to the filmmaking process that evolves into a rather amorphous sense of coauthorship. Coppola took unusual measures in every step of the production, including preproduction and post. Supervising sound editor David E. Stone (Plate 8) relates his initial meeting with Coppola in San Francisco prior to their first sound effects spotting session: "Francis just talked. He was not ready to show us the entire film. He discussed his artistic vision. He stressed the epistemological story and wanted the sound to be low-tech and organic. I reassured him that nothing would sound electronic." Coppola gave no specific instructions on the sound other than expressing his concern that it not sound synthetic, "that it would appear to 'come from the same place the story comes from,' and that it belonged in the world the character lived in."[10]

The beginning of *BSD* differs from the novel. Rather than start with the first journal entry of Jonathan Harker, Coppola begins his film with the backstory of Vlad the Impaler (Dracula), who fights the Turks in 1462 and comes home to Transylvania to find his beloved wife, Elisabeta, dead from her suicidal leap from the castle when she is misled to believe her husband has been killed. We then jump to 1897 where we see the descent into madness of Renfield, a lawyer recently returned from a visit to Transylvania. This results in Jonathan Harker taking Renfield's place to settle property transactions with Count Dracula. From this point onward, the narrative relays the story of Harker, Mina (Harker's fiancée), Lucy (Mina's best friend), Lucy's suitors, and Professor Van Helsing, including of their interactions with the bloodthirsty Dracula and his obsession with Mina, who resembles his dearest Elisabeta.

There is a fair amount of revisionism in Coppola's adaptation of *BSD*. Coppola did take some freedoms with the text that contributed to the development of Dracula's character, and added to the visual spectacle. As Vlad becomes Dracula and seeks vengeance, his discovery that his deceased wife, Elisabeta, has been pseudo-reincarnated in the form of Jonathan Harker's fiancé, Mina, is a departure from Stoker's original writing. The inclusion of the Cinematograph as a "Cinema of Attractions"[11] (which in Coppola's words is a "hall of wonders")[12] as a meeting spot for Mina and the young romantic prince (Dracula's alter ego) includes "nudie" shorts that were not entirely contemporaneous with the period.[13]

BSD is a feast of visual and aural stimulation. The entire style of the film recalls German Expressionism, gothic romance, art house eroticism, and Italian opera. Coppola wanted the musical score grounded in an Eastern European classical style. Rather than hire a well-known film composer,[14]

Coppola immersed himself in recordings of composers from Eastern Europe and decided on Polish composer Wojciech Kilar.[15] Kilar presented Coppola with three motifs for the film: the romantic theme for Dracula and Mina, the melodramatic theme for Dracula and his literal and figurative domains, and a third theme that served as an underscore to be used throughout the film. The two main themes served as leitmotifs. Coppola relied on music editor Katherine Quittner to edit the themes to better serve the narrative. Quittner had the original recordings with the separate tracks of the instruments, so she was able to vary the orchestration throughout the film.[16] Interpretive vocalises from Greek-American singer Diamanda Galas were edited in from CDs in several scenes to add a disembodied uncanniness.[17]

Sony, Columbia, and Coppola

In any contemporary Hollywood studio film production, there is a Kafkaesque logic to every project, which contributes to a sense that there is no particular entity accountable for what seems to be the faceless "other" who oversees the corporation.[18] This was particularly true when Coppola and Columbia/Sony agreed to make *BSD*.[19]

In 1989, Sony had bought Columbia Pictures and the MGM lot with the intention of acquiring Columbia's music and film content for Sony's thriving hardware and software business. Jon Peters and Peter Guber were ready to take Sony into a world of amusement parks, koi ponds, lavish productions, and, in short, to opulently outspend all other studios.[20] Coppola had a more than ten-year history with both Columbia and Sony. Coppola's failed history with Zoetrope Studios and its complicated distribution deals with Columbia, along with his penchant for attempting to leap forward with technology—often in its inchoate form—were infamous.[21] His famous economic "misses" of *One from the Heart* (1982), *Rumble Fish* (1983), *The Outsiders* (1983), and *The Cotton Club* (1984), measured next to his success with *Godfathers I, II,* and *III* (1972, 1974, 1990), alongside his bombastic epic *Apocalypse Now* (1979), made Coppola a risky, but possibly "profitable on a grand scale" director. With Peters and Guber at the helm of Columbia, the possibility of grand profits and excitement outweighed risk. Additionally, even with his checkered history with both Columbia and Sony, Coppola had a solid vision with *BSD*. He had agreed to shoot the entire film on the lot, and two young stars (Winona Ryder and Keanu Reeves) were now attached to the project.[22] Also, true to form—and an added treat to technologically friendly Sony—Coppola was going to be the first American director to experiment with Lightworks, a new digital picture editing system,[23] which was innovative and cost-efficient. Peters and Guber signed off on the Coppola project with enthusiasm.[24]

By the time Coppola was ready to begin postproduction on *BSD*, several dynamics had changed at Sony. First, Yoshiko Morita, the owner of Sony, had tired of the Peters/Guber excesses on the "film side" of Sony[25] and wanted to stop the spending, so Chairman Mark Canton moved Jonathan Dolgen (who had worked with Columbia previously and was recently hired by Sony) into the position of "budget tightener."[26] Second, Coppola was already overbudget and overtime in his production.[27] Third, Dolgen insisted that the postproduction sound be completed on the lot with a Los Angeles crew as a cost-saving measure.[28] These three events contributed to *BSD*'s descent down the rabbit hole of a technological mess, and one from which Hollywood and Bay Area sound professionals would emerge with a resulting collaborative sound design.

Who Is Going to Steer the Ship?

In spring 1992, David E. Stone was involved in the final mix of *Batman Returns*, along with Richard L. Anderson. Stone and Anderson were the supervising sound editors and were mixing on the Warner Hollywood dubbing stages[29] even as the Los Angeles riots in reaction to the Rodney King verdict erupted down the street. Stone had expected to work next on *Aladdin* with his colleague Mark Mangini;[30] however, circumstance had prevented that collaboration, so his immediate future employment was unclear.[31] Tom McCarthy, Jr.[32] (executive vice president of Theatrical and TV Sound Editorial at Sony Pictures) called Stone and asked, "How would you like to work on a monster movie?"[33] The movie was *BSD*.

From the very beginning, *BSD* was to be a "studio film." It was shot primarily on the Culver City studio lot. Coppola lives in California's Napa Valley and his studio, Zoetrope, is in San Francisco. Coppola had wanted Walter Murch to be responsible for the picture and sound of the film, and Richard Beggs, of both Saul Zaentz Pictures and Skywalker Sound, to be the primary rerecording mixer.[34] Murch was editing the *Godfather* trilogy,[35] and Beggs was still finishing his mixing duties with *Toys*. Coppola was insistent that the film be mixed in the Bay Area with mixers he was familiar with, but it became clear that other compromises would be made. The culture of sound editorial and mixing in Northern California and Hollywood were quite dissimilar. While Hollywood worked within a tight linear system of separated labor tasks and union duties, Northern California had developed a more fluid approach that encouraged crossing boundaries and sharing work duties.[36] Coppola had envisioned his preferred routine, in which Murch and Beggs would collaborate in their sound design and assist each other. However, with both men preoccupied, Columbia insisted on a studio sound job.[37]

The first issue was who would be supervising sound editor and who would mix the dub. Richard Beggs recommended Stone, who had worked on *Willow* (1988) and *Indiana Jones and the Last Crusade* (1989) at George Lucas's Skywalker Ranch, in Marin County,[38] to supervise.[39] A Hollywood A-List supervisor who had passed muster with the elite Skywalker group was a choice that Coppola could accept. The studio also wanted McCarthy to oversee the production, so both men were assigned the task of supervising the postproduction sound. McCarthy had been a supervising sound editor on many Columbia features, and when Sony bought Columbia, McCarthy was promoted to an executive position. He no longer supervised but wanted to cosupervise *this* film as a representative of the studio, and to have some control over its budget by using the studio's youthful—and less expensive—television editors.[40]

Coppola, with the support of coproducers Charles Mulvehill and Fred Fuchs, had insisted that the picture editorial be performed at Zoetrope and that the final mix be conducted at Coppola's winery in Napa.[41] Coppola wanted Leslie Shatz, a Bay Area sound editor and mixer, to head the final mix. Thus, a deal was struck: the film would premix (mixing many tracks of sounds down into fewer tracks for the final mix) on the Sony lot, and Shatz would head the final mix at Coppola's abode in Napa.[42]

Did Technology Drive the Decisions?

Traditionally, Hollywood sound editing had been performed on magnetic film for decades and the editors who worked in film had all undergone similar training for years starting as apprentice all the way to editor. Only a select few advanced to supervisory assignments. Thus, the responsibility of coordinating a sound editing team, spotting the film with the director for sound effects, auditioning and selecting sound effects from the sound library, assigning custom-designed effects, supervising the Foley, dialogue, automated dialogue replacement (ADR), and managing the schedule and costs were placed within the job description of supervising sound editor.[43]

Postproduction sound had only recently moved to 24-track tape rerecording on mixing stages, but Sony had transitioned to electronic sound editing for their television shows. Young professionals, skilled at the new technology but new to the film industry, were thrust into the role of sound editor for *BSD*. This executive decision caused anxiety among veteran sound editors who were still editing features on Moviolas and were already insecure about their futures at Sony, even though they were aware of their value as film artists. For Sony, under the frugal thumb of Dolgen, young editors who had edited some television on these early digital systems were a terrific cost-saving resource.[44] For Coppola, any venture into new technology could be sold as a positive.[45]

When Stone and McCarthy first met to discuss the division of their duties, it was decided that Stone would oversee the majority of editorial and all of the mix. McCarthy, who was wearing two hats—one as studio executive and one as union editor—would oversee the Foley, performed by Foley artists on the studio lot.[46] Stone would supervise the sound effects editors, who were young men with expertise in the music end of digital technology.[47]

Audioframe was a system utilized for sound design but unable to edit sounds, and Cyberframe was the PC-based editorial system modeled after the more familiar "editing bench" utilized in magnetic film editing.[48] Cyberframe was appropriate for editing preloaded library sound effects synchronized to the picture, but had no capacity for designing effects. Thus, the young men assigned to edit sound effects had to employ both the Audioframe and Cyberframe digital systems. They could not, however, interface between the two systems.

Stone brought in a veteran editor, Dave B. Cohn, to supervise dialogue and ADR. Cohn had been working on the Sony lot for a while by this time and had become an experienced Cyberframe editor. By empowering Cohn to oversee the ADR and dialogue, Stone could focus specifically on the challenging task of coordinating the multitechnological aspects of the custom-designed effects, as well as the library sound effects he would audition and assign to his young editing team. Cohn brought John Sisti (another Cyberframe editor) on board to assist with the ADR. Sisti had never edited on magnetic film and had come from a music engineer background. He was experienced with technology and could troubleshoot any issues with the Cyberframe workstations, in addition to editing.[49]

"We [Stone and McCarthy] made a decision that dialogue, Foley and ADR would roll [be edited] on Cyberframe," related Stone. "The [visual] wave took forever to build up on the screen, so we could turn off the wave form and watch the picture like we did with mag [magnetic film]."[50] Until now, Stone had been editing strictly on magnetic film and had not learned the new digital technology. So, he observed the young computer editors in order to learn the interface. When he believed they needed guidance toward a more cinematic approach to editing, he would mentor them as best he could without limiting their creativity so that they could understand what was unique about this particular film.[51] The trick was to empower the young talent, and to encourage their gifts, yet teach them the craft of filmmaking, which had been heretofore not been considered as a critical qualification by "Dolgenism."[52]

Stone then prepared to engage the sound effects library at Sony. What he found was disorienting, at best. Some of the effects were still in 35mm, which was not adaptable to any modern technology. Some of the effects had been transferred to DAT, at that time still a new technology. Some were on cassettes. Most of the recordings came from the 1950s MGM library and were not in the best condition for modern sound systems with enhanced clarity and dynamic range. To add to the confusion, the library notebook—then the official documentation of the effects—was not organized in any way that matched those sound effects that were actually available. It was standard practice by this time for studios and boutique sound shops to have a full-time librarian to log and categorize sound effects for editors. Sony, regrettably, lacked this resource.[53]

Collage of Collaboration

Stone decided to rely primarily on commercial CDs for a majority of the sound effects and hire a custom sound effects designer[54] for any special sounds he might require.[55] That task was assigned to Alan Howarth. Stone estimates that about 93 percent of the effects came from commercial CDs, and most of the rest came from Howarth, DAT copies of the sound effects library, special creative effects from his sound editors, and with some additional sounds from Leslie Shatz and Kim B. Christensen, who added synthesizer and MIDI-designed effects as surprise enhancements at the final mix.[56]

One of the decisions Stone needed to make was "who was going to do what tasks to what part of the story." He adds, "Rather than to let sound effects editors randomly design, we had to have a plan."[57] Three editors in particular, David F. Van Slyke, Sanford Ponder, and Christopher S. Aud, impressed Stone with their creativity and enthusiasm. The editors created special sounds and layered them in the Audioframe, with guidance from Stone. The sounds could be listened to, but not matched against any picture. The negative effect of this was that there was no visual referent at the time of design. The positive effect was that the editors' imaginations were liberated. They would then "output" the created sounds onto a DAT or optical magnetic disk (a 5" disk that looked like a CD) to be loaded into the Cyberframe. Then the editors could add these sounds to any of the more ordinary library effects available and combine and edit them in synchronization with the picture.

Ponder had previously published CDs of his own electronic music, so he was allowed maximum freedom with his own sonic images. One such image resulted in the sound of the gate to Dracula's castle as it closes behind

Jonathan Harker (Keanu Reeves) (Plate 10). Stone had found some magnetic film of common zoo animals from the 1930s and 1940s MGM recordings. He "auditioned"[58] them on a Moviola and copied them onto DAT. The DAT was then given to Ponder to use for the wolves and other animals that approach Harker in his carriage while on his way to the castle. Ponder designed a soundscape of the wolves in the background and incorporated the howls into the harsh and heavy clanging sound of the gate clamping down like teeth. The multilayered effect of the gate is a seamless transition from the real animals to the acousmatic animal sounds that enter the narrative.[59]

Howarth worked from home and owned all of the available design equipment. His responsibility was to design "signature sounds"[60] that were identifiable. One such leitmotif was the "wriggly whisper"[61] sound of the ancient Dracula whenever he floats. Another of Howarth's contributions was the special ambience of a "whooshing suctioning" for the magic gorge that Elisabeta (Winona Ryder) jumps into when she mistakenly believes her husband has died in battle.[62]

Since many of the sound effects were taken from commercial sound effects CDs, it was necessary to audition the CDs. However, the studio would not provide Stone or the editors with CD players. The solution was for each editor to bring his own personal consumer-quality CD player to his cutting room and listen to CDs there.[63] The professional-grade CD players run on a higher voltage and have a higher sound quality. Stone muses, "At one point, as I was looking at Sanford Ponder's CD player, I made the joke of how great it would be if some electronics company giant would buy a movie studio. All the electronic equipment needed would be provided. What a cool idea!"[64]

Stone illustrates the many layers of sound components that collaboratively complete the sound design of two scenes:[65]

> The BGs [background sound effects] and the mysterioso winds were designed by Howarth. The animal vocals are rough birds like crows, wolf effects and other animals. We never see Harker (Reeves) put his luggage on the ground. We're as disoriented as he is. The door slams (FX) [sound effects] on the carriage after Harker gets out. The driver shouts, which is production dialogue The stagecoach exits. We hear what may be a distant horse snorting, or perhaps a forest animal. It is deliberately ambiguous. There is thunder and lightning from FX. We zoom in on a wolf skull on a tree, and hear a scream-like design FX (Howarth or

Christiansen). Other layers of processed human vocals (by Diamanda Galas) and wolves were added by design and editorial. Leslie Shatz made them . . . in the mix. As the wolves gather, we layered quite a few actual wolf pack howls that were recorded at a distance. Since we could not record anything, we used some of the old studio recordings, along with commercial CD material. The close-up wolf growls are dog growl FX. The howls become overwhelming and omnipresent.[66]

As Stone relates this second collaboration, which leads to Harker walking up to Dracula's house:

When the carriage takes off, the woods fill up again with processed wolf pack howls, and other animals, including small birds that my young FX guys edited in. The carriage squeaks, shakes, and lamp creaks are all Foley. The horse whinnies are reverberated as they recede from the camera. Leslie's choice of reverb in the mix was appropriately unrealistic, and in this storybook context, plays well. We prelapped the gate [heard it before seeing it, as in Chion's acousmatic theory][67] so the audience wonders what the hideous metallic sound is before they see the spiked iron gate closing. It is supposed to be a version of a mechanical medieval monster mouth swallowing us up. Sanford Ponder managed to orchestrate dozens of sounds: slowed dumpster creaks, giant chain-driven mechanisms of all kinds, lions and tigers and bears (oh, my) and wolves, of course. I think we threw in every jail door, meat locker door, and ship hatch slam we could find into that moment.[68]

Revoicing, Revisiting, and Reediting

As Stone details every element of sound that is involved in this small section of a scene, it becomes clear how many components were required for the single effect of Dracula's gate. Cohn, meanwhile, was having his own brand of chaos. While it is not unusual for an ADR supervisor to travel to supervise an ADR session with an actor, Cohn traveled more for *BSD* than for any other project before or since. "At one time I had hotel room keys to London, New York, and San Francisco in my pocket," related Cohn.[69] There was far more ADR to be done in this film than was typical for two key reasons. First, Keanu Reeves needed to be completely rerecorded on an ADR stage to improve his line readings and accent.[70] Cohn recounts, "He was recorded at the Ranch."[71]

The other reason was Coppola's insistence that all visual effects be executed using techniques from the time period of the story (the silent film era), which required they be constructed on the set and shot in the camera, and were quite noisy. Cohn relates that the idea was great and the effects were wonderful. However, they added a lot of noise on the production track, which made production completely unusable for sound. So, all of the dialogue and sound had to be replaced. "It was great fun for the sound effects guys but murder for us," states Cohn.[72]

ADR and Foley were recorded on 24-track analog tape. Then the tracks were loaded into the Cyberframe for editing. Once the tracks were edited, it was necessary to "lay back" the edited tracks to 24-track tape for the rerecording mixing stage. The 24-track tape was then taken to the dub stage and loaded into the mixing board. Each editor was limited to one 24-track of effects, Foley, or ADR per reel. Dolby Laboratories had patented the noise-reduction card Dolby SR, which needed to be installed into the 24-track machine. Since track 24 was used to record the time code for synchronization and track 23 was left blank to prevent time code sound "bleeding" through to a recorded track, twenty-two Dolby SR cards had to be installed for each layback.[73] However, Mike Kohut, vice president of Sony's postproduction facilities, would not allow any of the remaining budget to be spent on Dolby noise-reduction cards. Sony was in development for its own noise-reduction system and Kohut refused to spend money for its competitor's product.[74] Other productions working on the lot were either independent of the studio and had their own Dolby cards or had obtained permission to rent the Dolby cards prior to this seemingly arbitrary change in policy. Kohut's decision was political, not logical, and this meant there would be no noise-reduction cards available, which for a feature of such a large number of sound effects tracks was problematic; the noise on each track would build up and would overwhelm the mix. Thus, the Dolby noise-reduction cards were a necessity.[75]

This was one of the largest obstacles the editors faced. They went around the Sony lot, and asked any editorial friends if they could borrow the SR cards for the layback. By borrowing a card here and a card there, they were able to manage the full 23 for each layback.[76] The sound was then encoded using the cards that reduced the noise. Once on the dub stage, they would be decoded back to the intended sound quality and dynamic range, minus the hiss of recorded noise on the tracks.[77]

After the premix, the tracks had to be transferred back to magnetic film. The purpose of this step was to allow the sound editors who would follow the final mix up to Coppola's winery in Napa to make any sound changes in the premixes on traditional editing benches. Coppola's facility was set up for editing sound on magnetic film and rerecording on 24-track tape. Thus, only experienced editors trained on "mag"[78] were qualified to make any changes and conform the sound premixes to the new version of the picture. Stone, Cohn, and Linda Folk, another ADR editor, went to Napa.[79] As is Coppola's method, the changes continued for the full six weeks of the final mix. Stone recalls, "We were making changes right up to the printmaster!"[80]

Music editor Katherine Quittner worked closely with Stone to carefully coordinate the amalgamation between music and sound effects. It is not the norm for sound and music teams to work in collaboration toward the resultant soundtrack. The more typical situation would have the composer and music editor working together, in tandem with the sound team during the final mix. In this case, however, since Coppola was helming the final mix at his residence in Napa, and Kilar had gone back to Poland, this left his master tracks under the careful stewardship of Quittner. Thus, she was left with the task of "repurposing"[81] the score as the film necessitated. Since the six-week final dub required all principals to be in attendance, Stone and Quittner were constant companions navigating themselves through a labyrinth of technological, artistic, and scheduling challenges. This provided a unique opportunity for Quittner to clearly observe the specific demands of sound editorial. She was able to deconstruct the tracks, with full knowledge of what the sound design entailed, and could contribute a score with varied orchestration and timbre. Coppola would give notes, and Quittner made the music changes, with full knowledge of, and even some minor collaboration *with* supervisor Stone. This was a unique experience for both Stone and Quittner.[82]

The resulting final mix of Coppola's *BSD* is complicated, rhythmic, and lush. The majestic style of the sound design, as crafted by Stone, his youthful and talented novices, along with Shatz and Christensen at the final mix, befits the effusive production design that appears on the screen. The unusual amount of ADR that Cohn supervised because of Reeves's unimpressive production dialogue and the noisy antiquated visual effects on the set created both crisis and opportunity for Stone's crew. Since most of the sound had to be designed, and there were significant budget and schedule constraints, the splendiferous soundscape is indeed masterful. While *BSD* was a coproduction between Columbia Pictures and American Zoetrope—with the corporate overlord of Sony Pictures controlling the finances—it was the collaborative innovation of the sound professionals that determined the creative outcome, utilizing incompatible technologies that they modified to accommodate their artistic pursuits.

Integrating the Shadow

This textual analysis including some of the specific aspects of the sound design Stone discusses in *BSD* concerns itself with the connection between Vlad the heroic warrior[83] and Dracula the terrifying vampire. The obvious denotative narrative connects the two into one entity: Vlad the warrior evolves into Dracula the undead predator as he rejects the God and Church he devotedly represented while he battled the Turks. The connotative narrative is one of the tragic hero[84] who is condemned to a purgatorial existence until he relinquishes his dominance over the woman he obsessively desires and allows himself to finally die.

In *BSD*, Dracula operates from two separate identities concurrently. One of these identities is a contrived and manufactured mask of controlled intimidation with cold-blooded intentionality and an obsession with self-preservation. This identity is in the persona of an ancient Count who seems to shuffle and glide his way around his mammoth and isolated castle with little to occupy his time. He changes his form and torments others to achieve his sinister goals. The second identity, however, is only alluded to with the appearance of his anthropomorphic Shadow that betrays the Count's inner emotions and psychology. The Shadow is, in actuality, Vlad, the fallen hero.

The character of the Shadow[85] embodies not only a visual identity that is distinct from the older Count but an aural one as well. This textual analysis will discuss the narrative importance of this alter ego of the Shadow of Count

Dracula, who demonstrates with visual cues and the aural design of Stone, Howarth, and Shatz, an archetype[86] of a deeper and more conflicted persona than that of the old Count, with whom the other characters more directly relate. Howarth's "wriggly whisper," the "backwards voices," and the whines and feminine sighs, all contribute to a sonic quality that adds power to the Shadow that visual rendering cannot accomplish.

The silhouette[87] displays the vulnerability of the human Dracula: the man who was capable of deep love for his wife Elisabeta. Although his appearances are infrequent, this Shadow dominates the scene. He expresses human desire and emotion. He is larger, more omnipresent than the Count. As the film progresses, the Shadow disappears as his human qualities slowly integrate into the Count until finally Dracula is fully one person: the man who loves Mina and wants to let her be free to be completely alive while he is free to be completely dead.

Before examining the interaction of the use of sound design with the Shadow in *BSD*, it is essential to introduce the initial experience of Jonathan Harker as he meets Count Dracula for it is in this introduction that we first encounter the Shadow. As Harker enters the Zeitgeist of the story world of Count Dracula[88] he begins the descent into the phantasmagoria that is Dracula's creation. Harker initially enters the domain of Dracula as a carriage pulls up on route to the castle and the driver's unnaturally long arm picks him up and places him inside the carriage.[89] The arm has an odd squiggly sound attached to it that alerts the viewer to an uncanny[90] quality to the driver (Plate 9). Harker's ride to Count Dracula's castle includes Harker relating to this driver that growls but does not speak, peering out the window and seeing the edge of the road that has a deep drop into the abyss, passing by a blue ring of fire[91] that appears out of the earth, and hearing haunting wolf howls and cries as he approaches the castle.

Harker is dropped off at the castle door and he proceeds up the stairs to the imposing edifice. We see Harker's shadow projected on the outside of the castle. It is unnaturally large but maps Harker's movements in a predictable way. The next shadow we see is of an image that projects on the wall of the foyer. Harker sees the shadow before the viewer does and registers some confusion about it. It floats in from the left and moves to the center of the wall. This is the introduction to the Count's Shadow[92] as a separate and anthropomorphic character (Plate 11). The Shadow has an accompanying unnerving leitmotif that is a combination of a whispery "whoosh," high-pitched whines, and a feminine sigh. The sound fades out as the Shadow does its pirouette and begins to turn away as we see Count Dracula in what appears to be full kabuki costume. The Shadow then stands fast behind the Count. The Count does not acknowledge the Shadow. Once the Count appears before him, Harker becomes focused on the appropriate introductions and dismisses any notice of the Shadow. As the Count introduces himself, "I am Dracula," another odd accompanying sound occurs. It is a combination of several voices played backward, as though his full complement of identities is introducing themselves along with the Count. Both the Shadow and the Count have identifying sounds that are separate and distinctive.

The Count is controlled, well mannered, and courteous. He invites Harker into the house and Harker steps over the Count's threshold. This action is signified aurally with a seemingly long exhale. Harker is now under the Count's control as well as subject to the whims of the related shadow. Harker has two villains to contend with, although he is, at this point, unaware of any conflicts.

Two scenes later, the silhouette is moving over a small picture of Mina. The leitmotif presents itself again. The Count and the Shadow are in tandem for a few moments. A femine sigh is followed by a deep growl—neither are audible to Harker. As the Count walks toward a map of London, situated directly in front of the viewer, behind Harker, the Count speaks, "I do so long to go through the crowded streets of your mighty London, to be in the midst of the whirl and rush of humanity [we then see only the Shadow and the Count's voice continues speaking from the same place in the room]—to share its life, its changes, its death." Harker turns to look toward the voice but sees only the Shadow. We hear the Shadow's leitmotif yet again as it floats off to the left, and the Count appears to the right, in front of the table. The voice that has been speaking to Harker has been disembodied.[93] It now emanates from the direction of the Shadow, not from the Count. Only after the Shadow leaves and Harker turns to see the Count does the voice reconnect with the personage of Dracula and disconnects from the Shadow. The Shadow now has a voice. It is the Shadow that has said, "to share its life, its changes, its death." The Shadow, not the Count, has spoken of life, change, and death.

Moments later, the Shadow has reappeared behind the Count and projected up on the map again. This time, however, its actions become more assertive. As the Count sees the small daguerreotype of Mina, the Shadow moves toward the photo before the Count and knocks over the figurine that holds ink and a pen. The ink spills onto the photo, looking like a shadow, as the Count picks it up and stares at it (Plate 12). Harker remarks that the photo is of his fiancée, Mina. The Count's leitmotif, accompanied by Gala's vocals, add a plaintive longing to his sadness. At this point, the Shadow turns toward Harker and reaches out to strangle him (Plate 13). The Count, trying to hold back his tears, remarks that he was married once, ages ago.

The Shadow now expresses the human emotion of jealousy and possessiveness that is understandable. Mina looks like Elisabeta and Harker is going to marry her. The Count remains calm and polite, not acknowledging any jealousy. Rather, he appears despondent. The two entities, the Count and the Shadow, are expressing two different emotional reactions to the same event. It is the Shadow that is unguardedly authentic in its response. There is no contrived behavior. It is the response of a man who sees another man as an obstacle between him and his wife. Here, the split in identities is clearer: the Count, old, regretful, and resigned to his fate, and the Shadow, angry, jealous, and possessive of his beloved.

The Count then insists that Harker stay for a month and to write his contacts in London to inform them of this. As the Count turns to leave he grabs hold of his cloak with his right hand. The Shadow follows him, but as it grabs the cloak with both hands, it lifts the cloak up like wings. The image of the bat—and the sound of the Count's cloak—comes from the Shadow, not from the Count.

As Dracula sets sail in an earth-filled box aboard the ship called the Demeter,[94] a Wolfen-like creature is loose and kills several mates aboard the ship. This Wolfen is another embodiment of Dracula, out to cause destruction. He appears at Lucy's home, screeching and howling for her. His growls and screeches are in eerie concert with both Gala's vocalise and a gothic sounding chorus. Additionally, Kilar's score builds with orchestration that features percussion. She goes to him in the torrential rain that has brought the

Demeter to London. The Wolfen violently rapes Lucy with an added element of eroticism as he fondles her and bites her neck. When he sees that Mina is watching the rape, he growls, "No, do not see me!" The Shadow is nowhere to be seen. Yet for one short moment, as Dracula implores Mina not to see him, there is a flash of the young Vlad imprinted on the Wolfen's face (Plate 14 and 15). He does not want his beloved to see what he has become. This is the beginning of an integration of the two entities.

When we next see Dracula, he is in the form of the young Prince Vlad. He is romantic, beautiful, and elegant. He sees Mina in London and clairvoyantly communicates to her, "See me. See me now" (Plate 16). She sees him but continues on her way (Plate 17). It should be noted that Kilar's film score transitions into a beautifully romantic leitmotif whenever we encounter the lovers.

As he tries to make her acquaintance, he is at first shunned by her. He successfully charms her and they enjoy the afternoon at the Cinematograph, viewing the new spectacle of moving pictures.[95] No longer simply interested in revenge and destruction, he is in ardent pursuit of his love interest.[96] Convinced he has a second chance to love and protect Elisabeta as Mina, he shows tenderness, warmth, compassion, and erotic romance.

While at the Cinematograph, Vlad coaxes Mina into a private area where he begins his seduction of her. The score transitions to the Count's theme as he speaks to her in his native language. As he begins to thrust his fangs into her throat, to ensure she will become like him and live forever with him, he stops himself. His passionate love for her keeps him from completing the act. What began sonically with his growl of lust ends abruptly as he recognizes the tenderness he feels.

Van Helsing and his crew of vampire slayers (Jonathan and Lucy's three suitors) proceed to destroy all the boxes of earth that have come from Romania on the Demeter. Dracula is watching, hanging from the ceiling rafters in the form of a demonic bat. He then transforms into a green mist (but sounds like a screeching bat) and travels quickly to the cell where Renfield resides. He kills Renfield because of his betrayal.[97] Dracula then proceeds to Mina's bedroom, and still in the form of a green mist, erotically slips under her sheets and appears as Prince Vlad. They begin to make love and she proclaims her desire to be with him forever. It is then that he tells her the truth of his existence. He confesses his identity and his crimes. She hits him repeatedly but then admits, "I love you. God forgive me, I do . . . I want to be what you are, see what you see, love what you love." He tells her she would have to die in order to achieve this goal. She assents and begins to drink his blood to live as he lives, but he stops her: "I love you too much to condemn you." She replies, "Take me away from this death." She drinks his blood against his objections, he reacts as if in orgasm, and they become forever attached.

As the vampire slayers enter the bedroom, Dracula, this time to protect his beloved, he transforms into a screeching demon. Dracula's voice is combined with bat vocals, and then treated with a harmonizer effect to enhance his intimidating demeanor. He and Van Helsing spar. Van Helsing reminds him that he had been an impaler, a murderer. Dracula defends his actions proclaiming that God has done this to him. He tells them that Mina is his bride and Harker shoots him. Dracula changes into a group of squealing rats to scurry and escape.

Dracula has one more transformation before his death. As the old and decrepit Count, he is dying after being stabbed and impaled by Van Helsing's crew. Mina has a rifle and is protecting the Count from the others. She helps him into his castle. Harker realizes he cannot force Mina to leave Dracula, "Let them go. Our work is finished here. Hers has just begun." Dracula lies dying at the foot of the Cross he had thrashed with his sword hundreds of years ago. He whispers to Mina, "It is finished." As Mina kisses the decaying Dracula he realizes she loves him dearly. The candles light, the scar in the stone Cross left from Vlad's impaling, heals miraculously. The old Count becomes the Young Vlad. He says to Mina, "Give me peace." She thrusts the sword in his heart deep and he dies, a final death that releases all under his spell. The thrust has two separate sonic qualities to amplify its importance: one thrust into the dying Dracula's body and one that connects onto the stone floor he lays upon. With this final act of love and compassion, Vlad has returned to his heroic incarnation with his beloved at his side. He dies an honorable death.

The transition from Dracula to Vlad the Tragic Hero requires redemption. The Shadow provides Dracula a way back to his true nature. Dracula has no interest in changing. He has chosen to split himself off from the noble hero who was victorious in the name of the Church. He lives a splintered existence. Dracula is capable of embodying many forms. He becomes an intensely erotic Wolfen, a huge mound of rats, green mist, a bat, and a demon (perhaps the Devil himself), and a gorgeously romantic young lover. Each incarnation is a manifestation of parts of this existentialist entity: none of them a complete representation of Dracula, rather only aspects of his splintered personality.

His Shadow operates differently. It fills the void of the human Vlad who was never gone, just split into two. As Dracula begins to change into the compassionate and loving husband, the Shadow disappears and integrates back into Dracula. With each encounter we have with the young and romantic Prince Vlad, we see the essence of who he is underneath his rage. Seeing an opportunity to reclaim his beloved wife in the form of Mina, Dracula, adept at changing his form and creative malicious horror everywhere loses his desire to do so. He remembers who he is. Try as he might to hang on to the blood-lusting vampire he has become, he is forever changed. He returns to his former self. The Shadow has served its purpose and exists no longer.

The sound design of the Shadow, from whispers and sighs, to growls, to squeals and vocalises, are all aspects of Vlad, the warrior. The inner emotions of the separated self, split into a carnal identity and a reflective one, which integrate back into the romantic prince, once he is resolved and has found his love, are expressed with visual effects and sound design. Yet without Stone and his team's aural contributions to assist in the specific delineations of events, the Shadow, as the emotional motivator for Dracula, is less clear. The meta-narrative of *BSD*—that of a broken man who becomes whole again—becomes more sensual and intuitive with the addition of the Shadow's subtext expressed aurally.

Notes

1   David E. Stone, Personal Interview, interview by Vanessa Theme Ament, Skype, in Savannah, GA, November 20, 2010.
2   Francis Ford Coppola and Kim Aubry, *Bram Stoker's Dracula* (Sony Pictures Home Entertainment, 2007).

3 Francis Ford Coppola and James V. Hart, *The Making of Bram Stoker's "Dracula,"* 1st ed. (1 in number line) (City: Pan Books, 1992), 6–9.
4 Ibid.
5 Coppola and Aubry, *Bram Stoker's Dracula*.
6 Coppola and Hart, *Making of Bram Stoker's "Dracula,"* 6–9.
7 His reputation is one of a perfectionist, but one open to all ideas. Those who work in the industry are quite open and frank about the culture of working with any and all directors.
8 "'The Blood Is the Life'—the Making of Dracula." Coppola and Aubry, *Bram Stoker's Dracula*.
9 The phrase "working without a net" is one I adopt to describe what is Coppola's methodology for collaboration. It denotes the work takes bold risks without knowing the outcome.
10 David E. Stone, Personal Interview (2010).
11 Referring to Tom Gunning's work: "Cinema of Attractions," In *Encyclopedia of Early Cinema*, ed. Richard Abel (London: Routledge, 2005), 124–7.
12 Coppola's narration over the feature film. Coppola and Aubry, *Bram Stoker's Dracula*.
13 Coppola has an interesting reflexivity in the scene at the Cinematograph. Although the films are slightly more recent than the film's periodization, it allows the viewer to see old films shot with the same kind of camera Coppola uses in the previous scene when Dracula as the young prince sees Mina on the London street.
14 Coppola preferred to have his father, composer Carmine Coppola, score his films, but the elder Coppola had died only a few months prior.
15 Kilar is also known for his piano expertise as the double for Adrian Brody in *The Pianist* (2002).
16 David E. Stone, Personal Interview (2010).
17 Ibid.
18 For information on corporate conglomerates and how they problematize filmmaking, a good text to read for a general overview is edited by Timothy Havens and Amanda D. Lotz, *Understanding Media Industries* (New York: Oxford University Press, 2012).
19 David E. Stone, Personal Interview (2010).
20 This was the gossip around the studio at the time and is confirmed in Nancy Griffin and Kim Masters, *Hit and Run: How Jon Peters and Peter Guber Took Sony for a Ride in Hollywood* (New York: Simon & Schuster, 1997).
21 Jon Lewis, *Whom God Wishes to Destroy . . .: Francis Coppola and the New Hollywood* (Durham, NC: Duke University Press, 1997).

22 Ryder had brought the script to Coppola because she had wanted to work with him on it. She had left *Godfather III* due to exhaustion and was hoping this project would provide the chance to collaborate. Coppola and Hart, *Making of Bram Stoker's "Dracula."*.
23 Coppola was the first American director to use Lightworks. It did not keep good sync and was another technological problem for the sound professionals. David E. Stone, Personal Interview (2010).
24 Griffin and Masters, *Hit and Run*.
25 Sony had purchased the music production side of Columbia as well, which was being managed properly and making excellent profits.
26 Griffin and Masters, *Hit and Run*.
27 David E. Stone, Personal Interview (2010).
28 Griffin and Masters, *Hit and Run*.
29 Warner Hollywood used to be the Samuel Goldwyn Studios. It has since become The Lot, another film company. Most recently, the Pickford Building was demolished and some of the remaining buildings are part of the Formosa Group, named for the famous restaurant across the street and contains sound editorial editing rooms and a mixing stage. Steve Lee, film librarian and historian, confirms this still holds true. Steve Lee, FB private message, November 9, 2019.
30 Stone and Mangini had known each other since their early days at Hanna-Barbera. Stone had been a sound editor when the young Mangini was a track reader. They both worked at Mangini's company (co-owned with Richard L. Anderson and Stephen Flick), Weddington Productions, on many pictures over the years.
31 An interesting anecdote: *Aladdin*, the Disney animated film, was nominated for the Academy Award for Sound Effects Editing alongside *Dracula*. Longtime friends and colleagues Mangini and Stone were both nominated that year for the award.
32 McCarthy was a supervising sound editor who had hired Stone to edit on several pictures for Columbia when the studio was merged with Warner Bros. as the Burbank Studios, and was located on the Warner Bros. lot.
33 David E. Stone, Personal Interview (2010).
34 Thomas McCarthy, Jr., Personal Interview, interview by Vanessa Theme Ament, November 15, 2012.
35 This was the collation of all three films into one continuous film in chronological time.
36 In my interviews with various Bay Area professionals from the San Francisco Bay Area, the general practice was that of collaboration and cross-pollination. The professional sound community was small and close-knit. Most of them had worked on *Apocalypse Now* and continued to work on various projects together. Their tasks varied. Walter

Murch encouraged work fluidity in others, as well as insisting upon it for himself. In my interview with Pat Jackson, she confirmed this by saying, "That's Murch."

37 Thomas McCarthy, Jr., Personal Interview.
38 The two communities had intermingled but are two distinct and different sound communities with different philosophies and approaches to postproduction sound. Walter Murch and Richard Beggs had worked with Lucas but were not part of the Lucas "house sound" that is identified with Skywalker. Murch and Beggs both worked with Saul Zaentz's Film Center in Berkeley and in New York as well. Both Murch and Beggs are often identified with Coppola and the Bay Area.
39 David E. Stone, Personal Interview (2010).
40 I had observed the addition of new "editors" from the music industry brought in and put to work on television shows. They worked at scale, whereas many veteran editors worked for substantially above scale. Several years later, I supervised Foley with some of these same editors. They were smart and enthusiastic but lacked the editing knowledge that the apprentice and assistance system had encouraged prior to the digital editing transition. Fortunately, they aspired to excellence, and learned from more experienced editors the nature of film craft.
41 The winery in 1992 is not the same winery imbibers enjoy today. This was Coppola's private home, where he began his entry into wine making, and was rudimentary in both its film apparatus and wine development.
42 David E. Stone, Personal Interview (2010)
43 Vanessa Theme Ament, *The Foley Grail: The Art of Performing Sound for Film, Games, and Animation* (Amsterdam; Boston: Focal Press/Elsevier, 2009), 35–43.
44 The conflicts of political economy mounted as studios valued saving money by hiring younger workers over more experienced workers, and the pressure cooker of veterans anxious about ambitious technophiles was brewing by the early 1990s.
45 Coppola's enthusiasm for engaging with new technology and innovation is well known in the industry.
46 This dual identity, as both an executive for the studio and a union editor on the film, made for some disruption for McCarthy. The two jobs were often in conflict with each other. When this was the case, according to both Stone and Coen, McCarthy chose the executive role as the determining one.
47 Cyberframe and WaveFrame are interchangeable terms for the same digital workstation. At this point, the Hollywood crews were calling the electronic editing Cyberframe. This system is the WaveFrame system developed by Chuck Grindstaff, as discussed in Chapter 2.
48 An editing bench was a table in an editing room that is used for making edits on magnetic film. While Moviolas were used for inserting the original effects, the bench was for conforming preedited effects.

49   David B. Cohn, Personal Interview, interview by Vanessa Theme Ament, Skype, in Simi Valley, CA, October 23, 2013.
50   David E. Stone, Personal Interview (2010).
51   Ibid.
52   This was the adopted term for the economics of John Dolgen, as he slashed and burned the budgets on various Sony film projects. Bottom-line costs, which hit hard during postproduction phase of *BSD*, were a direct result of Dolgen, who was not interested in the art, but only the cost.
53   Ibid.
54   Howarth is what is now known as a synthesist. He designs special sounds on synthesizing equipment.
55   Stone had to argue long and hard to get any budget for any custom effects for the film at all. David E. Stone, Personal Interview (2010).
56   Peter Sullivan is not officially credited in the film, but did, according to Stone, contribute effects for the thunder sounds in the film. It was not unusual for a sound professional to contribute a small part to a film and not be credited.
57   David E. Stone, Personal Interview (2010).
58   This process involves listening to many sounds, deciding on the sounds to be used, then having those recordings duplicated for editing use. This is more often how a supervisor would spend his or her time on a more typical soundtrack.
59   David E. Stone, Personal Interview (2010).
60   This is Stone's term for Howarth's leitmotif. David E. Stone, Personal Interview, interview by Vanessa Theme Ament, Chicago, IL, February 9, 2008.
61   The term for the description is mine.
62   David E. Stone, Personal Interview (2010).
63   Ibid.
64   Ibid.
65   Stone sent me a transcription of several scenes he recalled after examining his supervising notebooks, which he keeps for each film he supervises. In these notes, he included detailed accountings of who contributed which aspects of the sounds' authorship. I have edited the content for length purposes.
66   David E. Stone, "Sound Design Scene Descriptions From Dracula," November 20, 2010.
67   Michel Chion utilizes the term "acousmatic" to define a sound that we hear in a film, but which lacks a visual referent with which to attach its signification. We can normally identify the visual referent because of our familiarity with it, or because it has been shown to us previously in the film.
68   David E. Stone, Personal Interview (2010).
69   David B. Cohn, Personal Interview.

70 Reeves had offered to allow another actor to revoice his performance. He was disappointed in his work and wanted to allow Coppola to bring another actor in. Coppola insisted Reeves revoice his own character. The solution was to relax the actor, rather than to replace him. Both Cohn and Stone confirmed this with me; however, I was around the production at the time and recall the discussions myself. David E. Stone, Personal Interview (2008) and David B. Cohn, Personal Interview (2013).
71 David B. Cohn, Personal Interview.
72 Ibid.
73 Stone reminded me of this process during our interview in 2010. However, I watched the process firsthand during a visit to the lot in 1992. David E. Stone, Personal Interview (2010).
74 David B. Cohn, Personal Interview.
75 Ibid.
76 Stone, McCarthy, and Cohn all confirmed this fact. However, I also had firsthand knowledge of this, as I visited the lot during this time. David E. Stone, Personal Interview (2010), Tom McCarthy, Jr., Personal Interview (2012), and David B. Cohn, Personal Interview (2013).
77 The encoding/decoding process entailed encoding sound in a "stretched" or unnatural sound, that would then be decoded, or "normalized," for the best sound with the least amount of background noise or "hiss" and is not an issue in digital sound.
78 "mag" is the shortcut term for magnetic film.
79 This is another complication of mixing analog and digital. The changes had to be made on magnetic film by veteran editors experienced on the medium. Thus, Cohn, Folk, and Stone were charged with being in Napa "chasing mag" and continuing to keep up with Coppola's continuing picture changes up to the last day. While it is not uncommon for a director to make changes during the final mix, Coppola makes more changes than most.
80 Stone refers to Coppola's practice of making picture and sound changes up through the final mix, which is highly unusual. The final mix leads up to the printmaster that is the primary master of the film, from which all other prints are made. Some films would only have one, if the film was released in only one format. Some would have several, if it was printed in several versions for several forms of exhibition. It is time-consuming and elaborate. It requires a completely different technological setup for each type of exhibition release. Thus, Coppola's practice of changing the film to the very last moment puts tremendous pressure on all who are involved in postproduction.

81 The term "repurposing" now is more familiar, but at the time, this was not normative. A music editor did not typically take the composer's work, orchestrate each cue from the tracks, and essentially "score" the film herself as Quittner did. However, that was the task, as the composer was gone, Quittner was the lynchpin of the musical score, hired by Coppola, and an unexpected partner with Stone, as he related in his interviews.
82 David E. Stone, Personal Interview (2008). It should be noted that Quittner was unique as well: female music editor, working with a female picture editor (Anne Goursaud) and working with Stone (who had often put women in positions of authority when he was supervising sound editor), while all were working with a director who empowers rather than controls.
83 Coppola and Hart, *Making of Bram Stoker's "Dracula."*
84 Joseph Campbell, *The Power of Myth*, 15th ed. (New York: Anchor, 1991).
85 Janice Hocker Rushing and Thomas S. Frentz, *Projecting the Shadow: The Cyborg Hero in American Film*, 1st ed. (Chicago: University of Chicago Press, 1995)
86 C. G. Jung, *The Archetypes and the Collective Unconscious*, 2nd ed., trans. R. F. C. Hull (Princeton, NJ: Princeton University Press, 1981).
87 Coppola and Aubry, *Bram Stoker's Dracula.*
88 I refer to the older and predominant character by the moniker of Count Dracula or the Count. Other aspects of his persona will be referred to as the Prince for the younger romantic lover in pursuit of Mina and Vlad as the ancient Dracula who is Vlad the Impaler and is from the battle of 1462.
89 This moment in the film seems odd mostly because actor Keanu Reeves has essentially no reaction to the "grasp" and seems oblivious to the uncanniness of his surroundings. Whether this was an acting choice or a poor rendering of his character is open to debate. However, the moment arguably might have played better if he had exhibited some surprise at the event. Later, in Harker's journal in the book, he queries whether this driver was also Dracula in another incarnation.
90 Sigmund Freud, "The Uncanny," in *Literary Theory: An Anthology*, ed. Julie Rivkin and Michael Ryan, 2nd ed. (Malden, MA: Wiley-Blackwell, 2004), 418–31.
91 The circle, or Mandala, is a symbol in several religions that signifies a sacred space. This form appears several times in *BSD*.
92 From this point forward, Shadow is capitalized because it now is an anthropomorphic character and is distinguished from a pedestrian shadow.
93 This disembodied voice (one we hear but is not connected to a person on camera) is defined as acousmêtre by Michel Chion in *The Voice in Cinema* (New York: Columbia University Press, 1999).

94 Demeter was the Greek goddess of the harvest. She also presided over the sanctity of marriage, the sacred law, and the cycle of life and death.
95 They view "nudies" that were actually a few years later than the story, but the parallelism of the sexual content and Dracula's attempted seduction is delicious.
96 The concept of amor, a deeply romantic love separate from agape and eros, is discussed by Joseph Campbell in *The Power of Myth*.
97 Renfield has warned Mina to stay away from Dracula.

Plate 1 Supervising sound editor of *Barton Fink*, Skip Lievsay.

Plate 2 Barton Fink (John Turturro) as he rings the never-ending hotel bell in *Barton Fink*.

Plate 3 Barton Fink (John Turturro) enters the elevator, which incorporates Foley to emphasize its decrepit nature in *Barton Fink*.

Plate 4 The tight suction sound of Barton's (John Turturro) hotel room door adds to the sense of seclusion in Barton Fink.

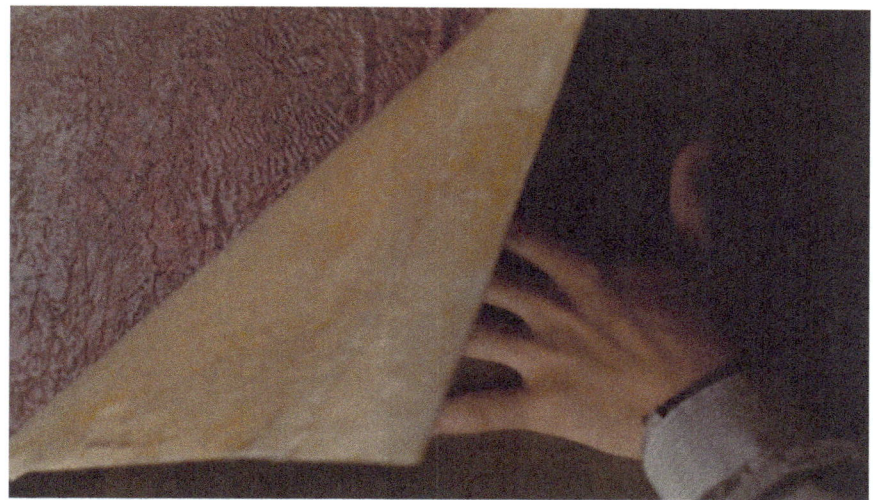

Plate 5 Barton Fink (John Turturro) attempts to reattach the dripping wallpaper in *Barton Fink*.

Plate 6 Charlie (John Goodman) and Barton (John Turturro) watch the wallpaper in Barton's room peel away from the wall in *Barton Fink*.

Plate 7 Studio head Jack Lipnick (Michael Lerner) bangs his desk, which results in the signature "hubcap" sound effect in *Barton Fink*.

Plate 8 Supervising sound editor of *Bram Stoker's Dracula*, David E. Stone.

Plate 9 The carriage driver's arm reaches out for Jonathan Harker (Keanu Reeves) in *Bram Stoker's Dracula*.

Plate 10 The Count's castle gate, which evokes sounds of animals and creatures, as it clangs shut in *Bram Stoker's Dracula*.

Plate 11 The Count's shadow travels around the room to greet Jonathan Harker (Keanu Reeves) in *Bram Stoker's Dracula*.

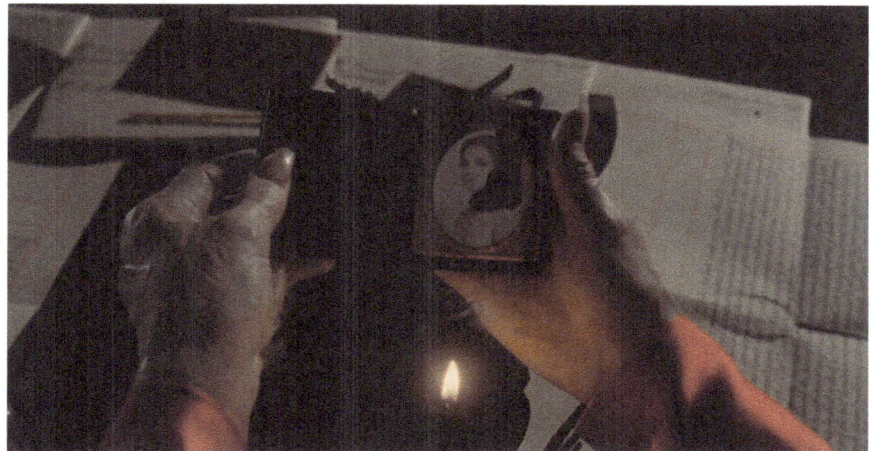

Plate 12 The Count (Gary Oldman) picks up Mina's photo, casting a shadow of ink onto it in *Bram Stoker's Dracula*.

Plate 13 The Count's shadow jealously attempts to strangle Jonathan Harker (Keanu Reeves) in *Bram Stoker's Dracula*.

Plate 14 Dracula, as the Wolfen (Gary Oldman), sees Mina (Winona Ryder) as he rapes Lucy (Sadie Frost) in *Bram Stoker's Dracula*.

Plate 15 As the Wolfen, the Count (Gary Oldman) appears as his young self and clairvoyantly communicates to Mina "Do not see me," in *Bram Stoker's Dracula*.

Plate 16 The Young Count (Gary Oldman) clairvoyantly communicates to Mina (Winona Ryder), "See me now," at the Cinematograph in *Bram Stoker's Dracula*.

Plate 17 Mina (Winona Ryder) sees the young Count (Gary Oldman) then continues on to the Cinematograph in *Bram Stoker's Dracula*.

Plate 18 Supervising sound editor of *The English Patient*, Pat Jackson.

Plate 19 Rerecording mixer of *The English Patient*, Mark Berger.

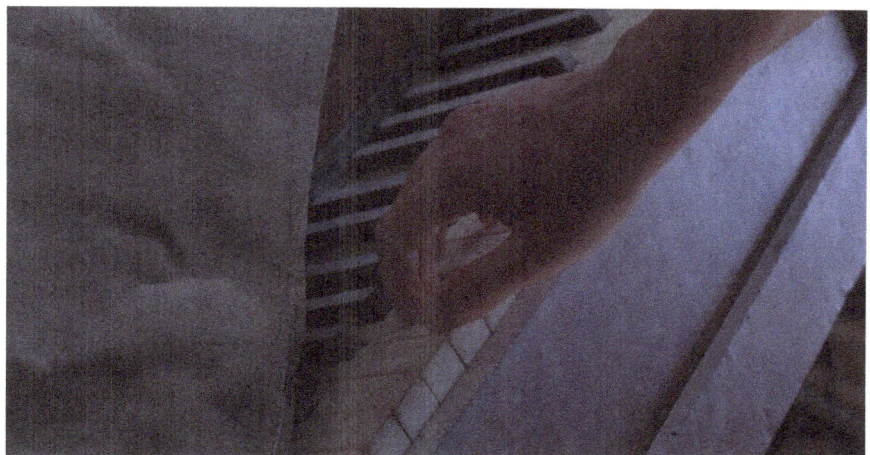

Plate 20 Hana (Juliette Binoche) begins to play Bach on the piano found covered in the monastery, in *The English Patient*.

Plate 21 Caravaggio (Willem Dafoe) taps the vial of morphine as Hana (Juliette Binoche) plays piano in *The English Patient*.

Plate 22 Kip (Naveen Andrews) discovers then defuses the bomb hidden in the piano: "Move that, and no more Bach," in *The English Patient*.

Plate 23 Katharine Clifton (Kristin Scott Thomas) counts the stars in the sky while Almásy (Ralph Fiennes) points to the sandstorm approaching, in *The English Patient*.

Plate 24 Members of the caravan rush to their jeeps during the sandstorm in *The English Patient*.

Plate 25 Katharine (Kristin Scott Thomas) and Almásy (Ralph Fiennes) discover romantic tenderness inside their jeep during the violence of the sandstorm in *The English Patient*.

Plate 26 Kip (Naveen Andrews) defuses a bomb while soldiers in military tanks approach in *The English Patient*.

Plate 27 Sergeant Hardy (Kevin Whately) tries to stop the tanks from crossing the bridge while Kip defuses the bomb in *The English Patient*.

5

"The Sound of the Desert Is Tape Hiss": A Study in Contrasts in *The English Patient*

> You can tell how quiet it is by what you can hear.[1]
> —Pat Jackson, supervising sound editor

When Pat Jackson (Plate 18) was asked by Walter Murch to take over the picture editing duties on *The English Patient* (1996) while he tended to his ill son, she did not know if she was prepared for the task. She was going to be the supervising sound editor, but for him to ask her to add the responsibility of helping to assemble the picture was, as Jackson expresses it, "typical Murch."[2] What Jackson means by this reference is that Murch does not acknowledge the more conventional boundaries between job descriptions normally assigned on a major motion picture. Rather, he has always preferred a more fluid approach that allows movement between the creative tasks, which began when he was a young man in his own artistic development.[3] As the key architect of the postproduction editorial and sound design process in San Francisco's feature film explosion of the 1970s and on, Walter Murch defines a clear and distinctive departure from both New York's single mixer paradigm and Hollywood's Fordist industrial studio methodology for designing and constructing a soundtrack. The filmmakers of the Bay Area are historically rooted in a less commercial and more "artistic" approach to the craft, which situated Murch perfectly to develop an approach to sound in film as a world apart from the two other primary American film cities.

The English Patient (*TEP*) is a study of how the San Francisco Bay Area in general, and Saul Zaentz Film Center—along with Murch's influence—more specifically, made the awkward transition of a major motion picture edited primarily using digital technology, that began as a conventionally edited analog film. The process then progressed to the complicated technological issues of several film sound technologies that both New York and Hollywood had grappled with years earlier. However, San Francisco started later, developed the processes more slowly, and made different technological choices after both Hollywood and New York had fairly well settled on adopting Pro Tools as its digital system of choice.

Walter Murch was in Italy conducting his normal film editing duties with the footage shot for the monastery scenes when he received the news about his son and decided to return to the United States. Murch agreed to return to the film later, but in his absence, Jackson would take over the general editing duties. This event changed the technological trajectory of the film as the delay offered opportunity to introduce new editorial technology. *The English Patient*, which won the Academy Award for Best Sound (an award for sound mix),[4] put Pat Jackson in the role as Murch's pinch hitter, in addition to her role as supervising sound editor, as she attempted to figure out "how to honor [Minghella's conceptual] aesthetic delicacy and distinction, and do it elegantly."[5]

The film is an adaptation from the book by the same name. Director Anthony Minghella wrote the screenplay of Michael Ondaatje's book, whose narrative was already visual and "filmic," so Ondaatje was not defensive about any changes made to his story.[6] While Ondaatje's book had been episodic and moved around achronologically, Minghella insisted the adaptation remain chronological in order for the story to hold together cinematically.[7] 20th Century Fox Studios originally agreed to produce the film, with Saul Zaentz as executive producer. However, while the production was shooting in Italy, Fox executives had decided they were unwilling to finance such a big-budget film with two English leads who were not major stars. The casting of Kristin Scott Thomas as Katherine was a particular area of disagreement, yet Minghella was insistent that she was the proper actress for the role. Fox executives wanted to replace Thomas with Demi Moore, and both Zaentz and Minghella held firm on Thomas for the lead. Fox pulled financial support for the film. While Zaentz went looking for additional financing, the cast continued to work on the film. The cast and crew deferred their salaries. Juliette Binoche continued to work and ate her lunch leftovers for dinner to save money while they looked for funding. When Miramax agreed to continue to finance the film, the budget had to be sharply reduced.[8]

TEP tells a fictionalized story of the real-life Count László de Almásy, a Hungarian cartographer and explorer with the Royal Geographical Society. While many of the events in the film are accurate, including the exploration of Egypt and Libya, the story includes an invented love affair between Almásy (Ralph Fiennes) and Katharine (Thomas), who is married to Geoffrey (Collin Firth), a wealthy Englishman helping to finance the expedition. The film focuses on Almásy and Katharine, with a secondary story that involves Hana (Binoche) and Kip (Naveen Andrews), an Indian mine defuser. Another character, Caravaggio (Willem Defoe) has known Almásy and suspects he is a spy (part of the true story of Almásy). The film involves two contrasting narratives: the desert where Katharine and Almásy are lovers and the Italian monastery where Hana nurses Almásy while he is dying of severe burns from a plane crash during an ill-fated attempt to rescue Katharine in the desert.

Conceptualizing and Executing

The two distinct worlds in the story, the desert and Italian monastery, were "a study in contrasts"[9] for both the visual and aural design. Originally, Minghella had wanted color distinctions—red for the desert and blue for the monastery—to be part of the production design. While not specifically executed, there was a residue of this concept kept for the film. One technique to keep the separateness of the two timelines was continual editorial jumps from one place to another. This was a challenge both for picture editing, as well as for the soundscape.

Murch had edited a large portion of the film on a flatbed.[10] Minghella decided that Murch's absence would allow time to make a transition to digital editing. Thus, the remaining scenes of the film would be edited digitally, but the footage that was already completed would need to be converted to the Avid digital editing system. Jackson was facing the first of many technological obstacles—as she refers to them, "the elephant[s] that went through the python"[11]—that she would navigate as she shepherded the film through the postproduction process, beginning with the first challenge: digitizing Murch's 35mm picture tracks.[12]

The analog picture was converted to digital on the Avid system—which Jackson had some experience with previously—and the remaining picture was edited digitally at Murch's Bolinas, California, barn on his two Avid systems. Jackson worked alongside Murch on the remaining footage, preparing general editing, as Jackson was not comfortable with the fine editing that is the mark of the more expert film editor.

Since this was the first major film that had been edited in the Bay Area on the Avid,[13] and computer memory was expensive, Murch was still experimenting with how to edit and incorporate sound and dialogue for the first rough edit.[14] In one scene, which had involved Almásy's friend discussing an event that was edited out of the scene, the scene was repurposed[15] with new dialogue shot in ADR. This alteration was made prior to the first preview of the entire film, which was also the first time the crew had seen the film in color. The Avid picture had been so unclear that the reedited dialogue was out of sync with the picture.[16] This was one of the major issues Jackson encountered with the disruption and transition from analog to digital technology in postproduction.

Those who work in the San Francisco Bay Area film industry frequently do not adhere to strict job definitions, while those in New York and Los Angeles more often do. Jackson, like many in the Bay Area, has worked on both documentary and narrative films, so she is accustomed to the flexibility required for both. This flexibility was particularly necessary on *TEP*. While Murch is credited as picture editor, Jackson worked alongside the master, and in addition to being credited as the supervising sound editor, she is also credited as associate editor. Jackson admits she got to see a movie edited together in a way most sound people do not: "I got to absorb the DNA of the movie."[17] The benefit of her expertise in both picture and sound editing was prescient in *TEP*, and later in the more developed digital era. Both Jackson and Murch utilized their technological adaptability from the late 1990s into the 2000s with greater ease than others who were more "fixed" as either picture or sound editors.[18]

"One of the advantages of working on all the digital computerized techniques that we have today, is that you have a great deal of flexibility and power in the sound area," Murch comments on the special features of the DVD of the film. "As we edit the film we can lay down the bones of what is going to be happening in the soundtrack."[19] One thing Murch did was turn the sound off while he edited the picture, so he could imagine what the soundscape should be: "Let me say that the sound is all in my head . . . I will make a couple of run-throughs and revisions with no sound at all."[20] Murch describes a process that is familiar to many sound designers: that of ignoring what is in the production track when the task is to go outside the ordinary world and add another imaginary dimension. Murch then describes his next step, "I'll talk to Pat Jackson, who is doing the sound design on the film, . . . and it begins to evolve."[21] As a final example of how Murch approached connecting the picture editing tasks with his role as sound collaborator with Jackson, Murch comments, "The monastery was a set . . . the floors were made out of painted wood . . . all we could use was the dialogue itself." "We have to then figure out what the whole soundscape of this film is going to be like, and that's something that I'm doing when I work."[22]

Generalist as Specialist

Pat Jackson, who has been teaching film sound at San Francisco State University, grew up a "military brat" and, as a young college student, had transferred to Stanford from the University of Kansas. Her major was communications with a specialization in broadcasting and film. So, while not a "born and bred" Bay Area native, Jackson qualifies as an enculturated Bay Area professional. Shortly after finishing college, she worked at the San Francisco public television station KQED. Documentary filmmaking was and still is a vital part of San Francisco film culture, and Jackson became an assistant working in 35mm documentaries. The film community in the Bay Area was then, and still is, small and close-knit. Jackson, along with others in the area, was hired to work on *The Conversation*. Jackson recalls the phone call to this day—from Coppola's Zoetrope office—asking her to work on the film synching dailies under the guidance of both Walter Murch and film editor Richard Chew.[23]

The nature of filmmaking in the San Francisco Bay Area is to do as Jackson did when she started out and "had a foot in both the documentary world and the narrative world."[24] She had recently supervised the sound on the documentary *The Celluloid Closet* (1995) when Jackson was asked to take on the supervising challenges that she would confront during the transition to digital editing on *TEP*. Jackson had also had experience on a digital system for sound editing. She had purchased her own WaveFrame system—the PC-based platform—for editing dialogue on *The Secret Garden* in 1993[25] (a Francis Ford Coppola production) and was less skittish about the transition than her colleagues had been in Hollywood several years earlier. Additionally, she had trained some of the sound editors at Skywalker how to use the WaveFrame.

Did Sonic Solutions Provide Solutions?

Saul Zaentz Film Center was slow to make the transition from analog to digital. A principal reason for this was economic. The duplication costs of magnetic film provided a primary source of revenue for postproduction houses. With the advent of a digital system, sound effects were accessible within the computer hard drive and thus, a main source of revenue would no longer be available. Roy Segal, vice president and manager of the Film Center, wanted to be sure that the digital system he chose for both Fantasy and the Film Center was likely to have longevity, before making such a costly investment. Rather than decide on a system precipitously, he waited to see how the systems fared in other facilities. Sonic Solutions had been developed by Jim Moorer, in the San Francisco Bay Area, and was in demand for cleaning and mastering music. Since Fantasy Records was a premiere Jazz record label, Segal saw the financial gains of purchasing Sonic Solutions. Segal, a former music recording engineer himself, was more inclined to follow the advice of his fellow engineers rather than the sound editors. In actuality, Segal never consulted any of his film sound professionals on the film side of the company.[26] Instead, he went with the expertise of his recording engineers, invested in Sonic Solutions, and insisted his sound editors all work on the new technology. By this time in the 1990s, most professionals were transitioning to Pro Tools, and other platforms were less frequently heralded as the future of sound editing and mixing. Yet, the transition to Pro Tools as the industry standard was not complete, and there were still hopes for other technologies to be adopted, so Segal was hopeful that Sonic Solutions would be viable for Fantasy/Zaentz.

One of the challenges that followed Jackson from production to the mix was a key decision of production sound mixer, Chris Newman, to use the new 20-bit digital Nagra system to record and mix the dialogue. Newman insisted on using a different microphone for each actor, and when Jackson pointed out that postproduction was not yet caught up to a compatible interface technologically, Newman brushed off her concerns. He believed he was providing her with the best sound possible and her job was to make it work.[27] What this exchange illustrates is how production sound and postproduction sound do not develop in parallel, nor do production mixers always understand how their tracks are used after they send them to postproduction.[28] This simple mismatch caused a critical problem for Jackson, and one that she masterfully overcame. Concerned with the consequences of editing dialogue from eight separate tracks of digitally recorded body microphones on the Sonic Solutions 16-bit system, which Jackson calls "the most editorially unfriendly platform I have ever worked on," she suggested a "bake-off"[29] audition for Murch, Berger, and herself to assess what would be the best sounding compromise. Her idea was to find a method to convert the Newman production tracks to viable tracks that translated to Sonic Solutions for the editors. Although the digital converter contained within the Sonic Solutions was adequate, she rented a superior outward converter, loaded a scene from Newman's recordings that contained broad dynamic range,[30] and converted it to digital 20-bit, 18-bit, and 16-bit. Her question for the three was, "Are we going to convert everything using the outward from 20 or truncate it down 4 bits to 16 to comply with Sonic Solutions?" One issue was cost. The outward converter would add a large expense to her budget if she needed to convert from 20 bits to more than 16. But if the sound would be superior, she was prepared to adjust her budget to reflect this necessity. They ran the scene and listened to the three choices for over an hour—20 bits, 18 bits, and 16 bits. Jackson reports, after carefully considering each contender, they realized they could discern differences, but, as Jackson states, "I could not perceive a 'best.'" The three decided to go with truncating Newman's digital dialogue at 16 bits and use the inboard converter, because it would require the fewest number of switches to be used when converting the tapes, which precluded errors in the transferring process, and was smoother and more consistent.[31]

By the late 1990s, Pro Tools was the dominant platform choice for sound editors. Since most of those who worked at Zaentz were freelancers, they had purchased their own Pro Tools workstations. There were some editors who had not been trained on the Sonic Solutions system, and one of these editors had edited his sound effects on his own Pro Tools system, then converted the edited tracks into Sonic Solutions. Jackson only discovered this fact during picture changes. When the changes in picture came, and all of his sound cues had to be conformed to match the new version of the picture, Jackson discovered original tracks were not in the Sonic Solutions. This meant that the editor had to make all of the changes on his Pro Tools, then convert them again into Sonic Solutions.[32]

Mark Berger (Plate 19), Walter Murch, and David Parker rerecorded the mix for *TEP* at the Saul Zaentz Film Center. Berger mixed the dialogue and some of the music, Parker mixed the sound effects and Foley, and Murch mixed the remainder of the music. The edited tracks from Sonic Solutions had been laid back to TASCAM hard disk recorders, and from that, the mixers rerecorded on to 6-track magnetic film for the final mix. While the editing had been a complicated recipe of Nagra digital production recording, Avid temporary sound recording, Pro Tools and Sonic Solutions sound editing, and 24-track analog Foley mixing, the final mix was completed on technology that was available from the 1980s.[33]

Foley Follies

The Foley for the film was recorded on 24-track magnetic tape, which was standard by 1996. However, what was not standard was having the Foley edited on Sonic Solutions. Segal's insistence on having all of the in-house editorial use his digital system added a great deal of work for the Foley editors. While Foley editor Malcolm Fife was accustomed to editing on magnetic film, Fairlight, and Pro Tools, Segal had waited a long time before allowing Fife to edit on Sonic Solutions. He found the process tedious. "Back then, hard disk space was scarce. We had to count every megabyte."[34] Many of the picture changes came while the Foley was being recorded. "It was very easy for Walter [Murch] to rebalance," says Fife about the picture changes on the Avid. "This was going on while we were recording Foley."[35] The Foley schedule was quite long, and Fife would have to decide whether to allow the Foley artists to finish a complete reel of Foley, and then do pickups later, or to do only partial reels, and insert the pickups after the picture changes were made. "In some cases, they [the Foley artists] wouldn't know the context. We transferred portions of the work in progress and then I would do a conform in Sonic Solutions. I would rewrite the cue sheets and fill in the holes with pick-ups. I remember the conforming quite well ... still have PTSD."[36]

During the first Foley playback for Murch, the Foley artists and Fife were anxious. "There were a lot of rules. He wanted the footsteps to sound random in quality, level, and synch."[37] Jackson recalls, "It was torture. Walter said, 'this isn't going to work. You can hear every footstep.' He wanted a 'specific range of frequencies. Have something else on the surface so you can have impact.'"[38] Fife recalls Murch instructing the Foley artists to "put broken glass down on the floor," so there would be lots of grit, but then adds, "they'd [the characters] be dancing on clean marble ballroom and we'd have grit!"[39] What resulted from this first encounter with Murch was a clear understanding that Foley would be "a huge test kitchen."[40] Fife adds, "I think it is a very respectful thing that he gave us all these notes."[41]

"Get Outside the Barrel of Yourself"

Aesthetically, *TEP* was approached as a true collaboration between director Minghella, composer Gabriel Yared, and the Murch/Jackson team to create a seamless soundscape with sound design and music working in orchestration with each other. Since Minghella wrote the screenplay as an adaptation from a book, he was mindful to keep the environments of the desert and the monastery distinct. It was his desire to have the sound—music and effects—work as a unit to create the magic of the aural background. In keeping with Minghella's ingenuity, Jackson kept notes in the margins of the novel as she read it.[42]

"Anthony's involvement with the soundtrack of the film is hugely deep and wide as far as the music goes. He and Gabriel have a way of working that is extremely intuitive and collaborative. They start coming up with musical themes early on in the film, sometimes before the film is finished filming." They send him themes without telling him where they will go to see what he might do with them. "Get outside the barrel of yourself. And that's what collaboration really does. What can somebody else bring that's unanticipated that makes the film bigger than the sum of its parts?"[43]

Jackson relates that "Anthony had Gabriel send sketches of music before the shooting started."[44] Berger adds, "Walter would use the sketches of what the music would be [while he edited]. They [Minghella, Murch, and Yared] would talk about dividing up the sonic spectrum—music to take care of this, frequencies, melody, sound, rhythm."[45] At times, the score was used as a sound bridge to assist the viewer with the flashbacks in the story. "Attention was paid to how the music starts and stops—weaving the music in and out of the desert," explains Berger.[46] Moreover, the sound design was used to create "a sonic memory from one place to another." To illustrate his point, Berger describes the scene when Kip and Hana are talking about Hana's cooking and Kip is pounding on a can of condensed milk—"tic, tic, tic," in the same rhythm as Almásy's coughing, which brings back the memory of the cobbler in the marketplace in Medina, and dissolves to the next scene. "It's like a flowchart," explains Berger.[47] Even the design of the desert wind, for Berger, was "a sonata for winds that has shape and a structure—a beginning, a middle, and an end. It had a lot of music concepts and was thought of in musical terms."[48]

The film provides opportunities for contrast: "The desert night is really, really quiet with an occasional clicking or buzzing, really localized," and the Italian military scenes contained "lots of clatter." Additionally, Jackson muses, "The sound of the desert is tape hiss." Jackson actually had traveled in the desert in Oman before for an industrial film project and remembered the sonic nature of the desert.

"One of my quarrels with action films is that you need pools of quiet that you can recharge your hearing with," Jackson adds. She wanted to ensure that the viewer understands that the monastery is a quiet place and that even the detail of the grass that Hana walks past when she walks up to the little pond contributes to the calm and peacefulness of the monastic environment. For this grass, Jackson recorded a California rattle grass that has little pods that shake when it moves. Jackson also points out the use of the quiet environs in the monastery when Hana first touches the piano keys as another opportunity for the resonance of the instrument to contrast to the stillness of the building. This attention to specificity is for Jackson an essential ingredient for a soundtrack befitting *TEP*.[49] "We paid a lot of attention to silence," adds Berger. "How do you create the sensation of no sound? You introduce something and then take it away. The taking away introduces a contrast that gives the impression of a crushing silence." Berger also credits Jackson for a detail that was instrumental for instilling a sense of stillness to the desert: "Pat was very creative in inventing bugs on a subconscious level."[50]

Berger provides of a different example of the use of contrast in *TEP* sound design as he describes the plane crash that ends Geoffrey's life and critically wounds Katharine. "It has to be detailed enough with sand flying into the propellers; big enough to sound real, but detailed enough to be convincing, and not so overwhelming that you don't believe she couldn't survive it."[51] What Berger chronicles is the contrast involved with detailed and specific elements that create the sense of chaos, while the titanic crash adds the emotive sense of doom, yet allows the hope of survival.

"Color Eats Sound"

One scene that Jackson describes, illustrates what she characterizes as "color eats sound."[52] Up through the 1990s, films were edited to black-and-white duplicated prints of film for economic purposes. Jackson recalls looking at the desert sandstorm scene and thinking, "This is just a wall of sound. It's just white noise." She was quite concerned that it was not clear and specific enough. However, she remembered that the "dirty dupes" that professionals used in editing were unclear and that the brain would attach the sounds more specifically to the objects in the scene once they were watching the scene in color. When she saw the same scene again, mixed to the color print, the massive editing of sounds for the sandstorm all attached to specific incidents on screen and was wonderfully effective. Jackson articulates this effectively when she states, "The human senses work together and inform each other."[53] Berger adds, "If you project a color image, the visual intensity overwhelms everything, and you really got a sense of how the sound got subsumed into the full image. The sound sort of blended into the picture."[54]

The book contains subplots that were left out of the screenplay, which left some story holes that needed to be repaired with dialogue cues. "When you lose whole subplots, you have to add lines to fill in dialogue to clue the audience in," says Jackson. One of Jackson's tasks was to meticulously go through the edited film and look for these holes and decide what dialogue needed to be included. One such addition was a line that explained Katharine's purchase of the thimble at the marketplace, which becomes a story point at her death. Jackson also reveals that some minor dialogue was added to include names for people, simply to help the audience identify minor characters visually.[55]

Jackson's small team of postproduction professionals managed to navigate through technological landmines while keeping Minghella's pure aural reverie in the forefront. The intimate nature of Bay Area filmmakers who know each other well, have worked together, are friends, and are loyal to the culture of the area influenced their interaction when decoding solutions to the myriad of technological challenges that arose during the postproduction of *TEP*.

Ephemeral Dualities

Count Almásy's languishing existence in an Italian monastery provides for him the rare opportunity to experience a truly existentialist transcendence. As he allows his morphine-induced disorientation to suspend him between two worlds, he teeters between his present with its physical pain, and his past with his lover Katharine and their emotional inconstancy. The elegance of Jackson's sound design, complemented with Berger's, Murch's, and Parker's pristine mix, brings to the dual soundscapes a pathos and delicacy that draws the viewer into his swirling mind as he remembers his passionate love affair, his attempts to save Katharine's life, and his struggles to avoid his own death.

Jackson, Murch, and Berger, all emphasize the dual soundscapes in *TEP*. This textual analysis looks at the use of contrasts and delves into the subtext of the quiet that surrounds moments of activity and the aural signifiers placed gently within the narrative to allow the two distinct soundscapes their own character. Just as Almásy experiences two separate incarnations—one as an ardent lover and the other a decaying soul—the aural narrative engenders two sentiments—the desert as isolation and the monastery as solitude.

Hana, Almásy's nurse, discovers a battered piano that invites her to take refuge from her duties and play one of her favorite Bach pieces. As she throws off the tarp covering the piano, the overwhelming weight of the disrobing creates a percussive and careless disturbance to the otherwise sedate environment. It is as if the piano has been in protective custody. Hana sits and the sounds of birds flow over her playing (Plate 20).

Meanwhile, in another room, while Hana continues to play, the sounds of Caravaggio preparing his injection of morphine (an addiction he has acquired from the relief of his own injuries) include the liquid spritzing out of the top of the needle, the tapping of the syringe (Plate 21). "I've come to love that little tap of the fingernail against the syringe ... tap ... tap ... tap ... ," says Almásy, and the stretching of the rubber band around Carravaggio's arm. As Carravaggio breathes more deeply, the piano transitions into an allegro section of the Bach.

Kip, the mine defuser for the British Army, runs into the room instructing Hana to stop playing. He informs her that the Germans were clever at hiding mines in unusual places, including pianos. Alas, he finds a device attached to Hana's piano. The juxtaposition of the music of the piano and the birds against the danger of the mine and the drug addiction illustrates not only the power of contrasts within the story but also the underlying emotion contained within the sounds of Bach and singing birds while potential death from a drug overdose or explosion is a daily possibility. Hana's calm and humorous demeanor contrasts Kip's vigilance in his detonation. "Look. See? Move that, and no more Bach," says Kip as he discovers the mine (Plate 22).

In another scene, as Hana bathes Almásy and soothes his skin with wet cloths, the stillness of the room is interrupted only by two sounds that accompany their dialogue: his labored breathing and the cloths as they enter and leave the water bowl. The very next scene has Hana sitting in her garden with two main sounds, yet again: the musical tinkles of her handcrafted wind chimes of glass and shells and the motorcycles that pass her, one of which Kip is riding. The silence that encompasses the simplicity of dualities—one harsh, one soft—is a repeated trope[56] in the film.

The sound of the motorcycles segue into one of Berger's sonic memories as Almásy lays in his bed, musically chanting, and the motorcycles take him back in time to the Sahara as he and his colleagues are in Jeeps crossing the desert. The memory evoked is that of him and Katharine in one of their early encounters. The principle sound is that of their jiggling Jeep, dictating the rough and unfriendly desert terrain, as the two engage in adversarial banter. She and the rider atop the Jeep begin singing, which acts as another sound bridge to a close-up of one of the Muslim guides chanting the prayers for one of the daily bows toward Mecca amidst the expedition—the same chant that had prompted the enfeebled Almásy's memory. This circular aurality—motorcycles, Jeeps, chanting, song, chanting—is surrounded with a vacancy of sound. Negative space, again, is a key component of the sonic canvass.

As Katharine sits, smoking her cigarette, planted gently on the desert sand, others are packing equipment in the background. Only crickets can be heard above the packing. She remarks she is counting the stars and rearranging them up in the sky. Almásy asks her to look at the oncoming sandstorm, which will make its way toward them quite soon, "In a few minutes, there'll be no stars. The air is filling with sand" (Plate 23). The stillness is shaken with the cut to the ferocious storm, an orchestration of sand and winds, all varying in speed, sizes, shapes, and pitches. The storm obfuscates the view, so the sound drives the narrative. Dialogue pokes through to indicate the principals are rushing to get into their jeeps with their valuables (Plate 24). Occasional Jeep door slams interrupt the onslaught of ringing wind and accosting sand. Katharine and Almásy sit again in the same Jeep, but this time, the sandstorm is an opus that allows the impending lovers to be isolated. The impassioned winds swirl around them erotically and imprison them—allowing no escape—present the possibility of inevitable demise. The Count describes, poetically, various types of desert winds, and delicately touches her hair. Violence outside their cocoon begets tenderness inside (Plate 25).

Almásy's initial discovery of the Cave of Swimmers is a lonely one. While the Muslim prayers continue, he walks into the cave unaware of what it contains. The interior of the cave reflects his solitary footsteps and his initial shouts of discovery. As the others join him, these reflections amplify the quiet utterance, "My God, their swimming," as Peter Madox realizes the significance of the drawings on the cave walls. The stillness of the moment enhances the importance of the statement: in the desert, the natives had a body of water in which to swim. This cave is his momentous discovery and later will become the demise of his lover's hope.

At another expedition to the cave, Almásy is alerted to the sound of a small plane approaching. Katharine's husband, Geoffrey, under the guise of offering to pick Almásy up, but jealous at the discovery that Almásy and his wife are lovers, deliberately flies his plane toward the Count in an effort to kill the three of them (Katharine is in the plane as well). The sound of the plane's angry aim toward Almásy alerts him and he leaps out of the way. The sonic shredding of the plane as it barrels past him is devastating. Pieces of metal of various sizes and sound pitches fly. Geoffrey is killed, and Katharine is seriously injured.[57]

As he carries her from the plane, the film score encases the soundtrack. This is the moment when Katharine states, "I've always loved you," as she reveals that she keeps the thimble he bought her and wears it as a necklace at all times. He lays her down in the cave and this wonderful place of discovery will now become her tomb. The reflections inside the tomb enhance the fire he has constructed for her, and the bits of wood he still breaks to keep her warm, "I don't want to die in the desert," she proclaims, almost knowingly. He promises to come back for her with help. While the score does swell with the predictable full orchestration, their final kiss is audible, tender, and simple.

Hana and Kip have their first romantic evening together after several encounters at the monastery. One night, Hana walks outside the main house to discover burning flames inside shells lighting the way to another building where Kip sleeps, like luminarias. She follows them filled with delightful anticipation. The night is still, with only her footsteps to accompany the delicate flickering of the flames. The underscore contains light orchestration featuring the piano, as if played by Hana, recalling their initial encounter. As she enters the building, only Kip's voice disturbs the stillness, "Hana," to which her smiling reply is, "Kip." The two begin their romance, which is complicated by cultural differences. His experiences as a Sikh raised in England, and hers as a privileged Canadian, bring different perspectives on the war. This plot point is more developed in the novel.[58]

Hana has lost several people that she loves in the story and is concerned that Kip will be another casualty. When he is asked to defuse a bomb in a well, she gets anxious. Kip's isolation in the well, with the specificity of the sound elements for the detachment of the bomb components, adds tension to the scene. The spraying of the gaseous oxygen, the specific hitting of the metal tools against the bomb, and the delicate water drips and motion in the well are juxtaposed against the raucous approach of soldiers in military tanks nearby. Hana's bicycle provides a sonic connection to the two disparate scenes. She rides, the soldiers rambunctiously celebrate, and Kip meticulously attends to his craft, aurally building tension to some kind of climax that has yet to occur (Plate 26). The rumble of the tanks is sensed with a low-end vibration in the well, and Kip feels the pressure to complete his task. The auditory storytelling here is discrete and precise. The tools begin to be affected by the vibration and move around his workspace. Some fall into the water. His friend, Sergeant Hardy tries to stop the tanks from crossing the bridge right above where Kip is working, lest the vibration detonates the bomb (Plate 27). The footsteps of British soldiers running across the bridge above signal their desperation to stop the vehicles. Just as the increasingly loud tanks approach, Kip has to choose which wire to cut. The tanks stop, the sharp cut of a wire is heard, and the bomb is diffused. As Kip comes up from his "fog of war,"[59] what is discovered is the reason for the exuberance of the soldiers: the war is over. The irony is apparent. The sounds are not realistically balanced but are specifically chosen for plot and drama. The sound design tells the story, and no dialogue is needed, although it does indeed exist.

This celebration continues into the evening, with Kip's friend Sergeant Hardy leading the way. Dressed only in his boxer shorts, he drunkenly climbs up onto a statue to proclaim victory. As the scene cuts to Kip celebrating in

Almásy's bedroom with Hana and her record player, the distant sound of a bomb is heard. Kip turns toward the sound.

Ever the disciplined bomb diffuser, he darts off. Although the sound is far off, the quick cut from his best friend Hardy is an uncanny[60] foreshadowing. With only some scant dialogue indicating it is indeed Hardy that has been killed, Kip drives up on his motorcycle—the only sound as he approaches—as the celebrating is over, most of the people have left, and Kip's only friend, except for Hana, is gone. Hardy's fiancée remains, as does another British soldier, explaining that the statue was "booby-trapped." Hardy is in an ambulance, covered because he is unrecognizable.

In the next scene, Kip sits on his bed, with birds singing, and Hana pleading to come in and talk to him, as she bangs, pounds, and kicks his wooden doors. A melancholy piano plays. This recurring piano as a leitmotif that signifies the relationship between Kip and Hana has changed, as clearly, so has Kip.[61]

Kip's departure is introduced with the disassembling of his tent—full screen and loud. His domicile is gone. "We've been posted; north of France," he tells Hana. "The patient and Hardy, they're everything that's good about England." The cloth that Kip folds as he is collecting his belongings dominates the soundtrack, accompanied by birds and the rattle grass that Jackson so carefully included in the monastery aural environs.

Kip's leave-taking, regrettable though it is, bears sharp contrast to the disconsolate rupture Almásy experiences as he returns to the cave where Katharine had recently painted likenesses of the "swimmers." The beginning of the film, which comes full circle as he finds her lifeless body and attempts to take her back to England so she does not have to "die in the desert," brings the two aspects of Almásy's life into focus. As he entreats Hana to let him die from a morphine overdose, the delicate beauty of the glass containers is cut open while accompanied by the piano leitmotif. The overdose allows him to die peacefully as Kip, Katharine, Caravaggio, and eventually Hana all leave.

The constructed narrative of *TEP*, set in two geographies, encompassing two love stories, and even examining two incarnations of one man, required painstaking and meticulous detail in the design of the two soundscapes. The collaboration between music and sound design—both diegetic and nondiegetic—allowed delicate moments to be heightened aurally, allowing the focus to pivot toward the delicate, the unobvious. Jackson's sound design, with its specificity and detail, highlighted key story points that would could have been easily lost in such a subtle and nuanced narrative.

Notes

1. Pat Jackson, Personal Interview, interview by Vanessa Theme Ament, Skype, in San Franciso, CA, July 18, 2013.
2. Ibid.
3. In *Making Waves: The Art of Cinematic Sound* (2019), Murch's process is illuminated by him and others who worked with him.
4. Source: The Academy of Motion Picture Arts and Sciences, 1996, Miramax Pictures.
5. Jackson maintained a close relationship with both Murch and Minghella throughout the postproduction process. The sound design was one of collaboration and consensus. Pat Jackson, Personal Interview.
6. Anthony Minghella et al., *The English Patient* (Burbank, CCA: Miramax Home Entertainment, distributed by Buena Vista Home Video, 1998), Special Features, Michael Ondaatje.
7. Ibid.
8. Pat Jackson, Personal Interview.
9. Ibid.
10. A flatbed is an editing machine that was used for magnetic film editing. It allowed the editor to watch the footage with the reels laying flat and moving from left to right.
11. Pat Jackson, Personal Interview.
12. Ibid.
13. Malcolm Fife, Personal Interview, interviewed by Vanessa Theme Ament, Phone, San Francisco, June 1, 2014.
14. Feature films have a first edit, the "rough edit," which is the preliminary edit that postproduction works from, with the director's permission. Other edits follow that include more refinement and delete or add scenes.
15. The original intent of the scene was altered, so the "repurpose" of the scene required dialogue to align with the new intention of the narrative.
16. Pat Jackson, Personal Interview.
17. Ibid.

18 Some sound editors took Motion Picture Editors Guild training on various digital systems, only to find the standard in the industry changed from one platform to another.
19 Minghella et al., *The English Patient*, Special Features, Walter Murch.
20 Ibid.
21 Ibid.
22 Pat Jackson, Personal Interview.
23 Ibid.
24 Ibid.
25 Ibid.
26 Malcolm Fife, Personal Interview (2014).
27 Pat Jackson, Personal Interview.
28 Jackson relates how years later, Newman sat on a rerecording stage and was amazed at how dialogue was mixed after being edited. He expressed to Jackson a new sense of appreciation for dialogue editors. This is a key problem with the separation of production and postproduction. While postproduction professionals are aware of the consequences of production sound mixing on their work, the reverse is not as often true. Pat Jackson, Personal Interview.
29 This term refers to the traditional bake-off the Academy of Motion Picture Arts and Sciences that was traditionally held each year to decide the nominees for best sound editing.
30 Jackson was quick to point out that the suggestion to select a scene of broad dynamic range was mixer Berger's idea. One quality I noticed that Jackson, Berger, and Fife share is the generosity to give credit where it is due. Pat Jackson, Personal Interview.
31 Pat Jackson, Personal Interview.
32 Ibid.
33 It is important to note that *TEP* is the third film to be included as a case study and was in postproduction in 1996. By this point, the majority of films were edited digitally, yet the transition was still incomplete, and a great deal of disruption still existed.
34 Malcolm Fife, Personal Interview (2014).
35 Ibid.
36 Ibid.
37 Pat Jackson, Personal Interview.
38 Ibid.
39 Ibid.
40 Pat Jackson, Personal Interview.

41 Malcolm Fife, Personal Interview (2014).
42 This is not a practice I had heard of prior to Jackson's interview. Hollywood sound editors have a work process that negates any time for this element to be included. It is possible to get a script, but the workflow is more standardized. It would be highly unusual for a Hollywood sound designer to be on a film early enough to read the book from which an adaptation developed.
43 Minghella et al., *The English Patient*, Special Features, Walter Murch.
44 Pat Jackson, Personal Interview.
45 Mark Berger, Personal Interview, interview by Vanessa Theme Ament, Skype, in Berkeley, CA, July 17, 2013.
46 Ibid.
47 Ibid.
48 Ibid.
49 Ibid.
50 Mark Berger, Personal Interview.
51 Ibid.
52 Jackson states that this is a concept she and her Bay Area colleagues would comment on when editing sound on various films. I mentioned to her that in all my years of working in Hollywood, I had never heard anyone analytically discuss this phenomenon. There is a distinct intellectualism toward filmmaking that occurs in the Bay Area that is less often the regular day-to-day conversation among sound professionals in Hollywood. Rather, in Hollywood, the proverb would be, "it sounds better in color," followed with a laugh. Pat Jackson, Personal Interview.
53 Pat Jackson, Personal Interview.
54 I actually interviewed Berger before Jackson, and they discussed this same phenomenon. It is interesting that they both remarked on this without any prompting. Mark Berger, Personal Interview.
55 Pat Jackson, Personal Interview.
56 This trope, or motif, is of simple sounds that contrast surrounded by quiet or stillness.
57 This is the scene Mark Berger refers to earlier as "big enough to sound real, but detailed enough to be convincing, and not so overwhelming that you don't believe she couldn't survive it."
58 Jackson states that scenes involving their relationship were cut from the film, which diminishes the magnitude of his leaving near the end of the film. Pat Jackson, Personal Interview.
59 I am alluding to the documentary *The Fog of War: Eleven Lessons from the Life of Robert McNamara* (2003). Here, I mean it to imply the altered realities that come from the constant barrage of fear and stress that accompany battle. More than being

shell-shocked, Kip does not fully understand what is happening and he interprets his situation inaccurately.

60 The sense of uncanny here is the device in film that signals the contrast of the celebrating and the bomb exploding having a connection that is yet to be revealed. Sigmund Freud et al., *The Uncanny* (New York: Penguin Classics, 2003).

61 In the novel, it is the dropping of the bomb on Hiroshima that turns Kip's heart. As a loyal British subject, he had ignored the racism so prevalent in his country. His brother, on the other hand, had not. They were not politically in agreement. After the United States bombs Japan, Kip realizes he does not belong to the dominant culture and he leaves. The film alters this plotline. Michael Ondaatje, *The English Patient* (New York: Vintage Books, 1993).

6

Conclusion: Reassessing Sound Design as a Collective Endeavor

A Cultural Transition

The transition into digital sound editing is partially a technological story, but not entirely so. To present the piece of the puzzle from a more cultural history, framed within the specific geography of three large and critically important American film communities, allows us to broaden our view of how film sound editing and mixing developed as nascent digital creative crafts from the previous analog technologies and labor practices. Scholarship's grand narrative has often portrayed sound design as a specialized pursuit of a few elites, and the rest of postproduction sound "workers" as technicians who execute the "art" of the masters. This is a false concept based, in large part, on a lack of understanding of the craft itself, and a misconception about the postproduction sound community in general. It is easy to valorize those who are gifted at self-promotion or have been fortunate enough to work with a high-profile director or producer. The true depth of film sound practices is more richly mined when including the experiences of those who are less known, perhaps, but who have worked within the culture enough to reflect the history with proper context. As is true with any developing discipline in academia, sound studies and production studies are both adolescents, and there is plenty of room for reassessment of previous assumptions. What I hope to have contributed with this small addition to the compendium is some evidence that postproduction sound professionals in films are creative, artistic, and imaginative agents who work independently, as all artists must, and who also collaborate as all filmmakers do.

The three communities, New York, Hollywood, and the San Francisco Bay Area, had distinctly different approaches to the transition to digital sound editing for feature films. New York—the first to make the transition—chose a platform related to an already established technology that was manageable for the few professionals who used it. The crews were small and had worked together previously for Ethan and Joel Coen. The supervisor, Lievsay, owned the company and controlled expenses and had input into how to best exploit the technology to make it financially worthwhile. For Lievsay, leaving analog behind meant having more control over his sound design, and less reliance on the whims of a single rerecording mixer who might disagree with his sound editing choices, yet who would be victorious at the final mix. Simply put, Lievsay gained freedom by leaping into the new technology as soon as possible. Whatever bumps he or his small crew might endure for the sake of this creative freedom seemed a small price to pay for the liberation of their ideas and the possibility of expanding their skill sets.

Hollywood was also quick to enter the transition, but the Taylorism[1] of labor practices, the differences in skills and schedules that divided television and feature sound professionals, the alienation from the mammoth corporate structure of Sony regarding the needs of the editors, and the simple fact that the technologies were not developed well enough to interface with each other created havoc for the talented sound crew. Due to the creative requirements of the sound design, Stone's indefatigable persistence to reconcile his youthful team's exuberance with the solid Fordist practices of the older union professionals, and the seemingly insurmountable technological challenges, Stone and his colleagues were galvanized to meet each obstacle with enterprising innovation and humor.

The San Francisco Bay Area made the transition to digital editing with a later, more relaxed, and more experimental process. The Saul Zaentz Film Center made a technological transition only after much careful deliberation, and, as it turns out, the selection was in conflict with what would become the more accepted standard platform in a few years—Pro Tools—in both Hollywood and New York. Because the Film Center also recorded music and wanted the investment in new technology to "serve two masters," Roy Segal chose a technology that was excellent for music recording sessions, but less well-suited for the needs of feature films. Thus, this experimental approach—an innovative strength for both the Bay Area film culture and the Film Center—provided several key challenges for Jackson and her crew. However, the focus on the aesthetics, and the ability to work slowly and carefully, allowed Jackson to keep her "eyes on the prize" and allow the sound design to evolve as Murch kept reediting the picture. The sound editors' experience with different platforms by 1996 informed their abilities to manipulate this limited technology to do what they needed. Jackson's sense of detail, her devotion to director Minghella's ideas, and her trust in both Murch and her crew allowed a sense of collaboration to unfold during the generous postproduction sound schedule.

Exploring the cultural geographic lens through industrial ethnographic practices, which is so important for properly situating and contrasting each film community, allows a clearer analysis of the distinctions between the three sets of film professionals. First, New York had substantially smaller crews and shorter schedules for editing and mixing. The convention of one rerecording mixer—as established by Dick Vorisek—is still the normal practice in New York. In the 1990s, Hollywood utilized the three rerecording mixer model, and the Bay Area followed that model as well. Crews in the Bay Area were larger than New York, but smaller than Hollywood, even on Zaentz's lavishly produced films. The schedules were longest in the Bay Area.

In both Hollywood and the Bay Area, two Foley artists were usually employed on a major film. There were times in the 1980s and 1990s when the Film Center would employ only one.[2] However, New York has always followed the practice of employing one Foley artist, except for very unusual circumstances.[3] In both the Bay Area and Hollywood, there is at least one dedicated Foley editor, and on major features, often more. However, in New York, editors are more likely to share editing duties. The job descriptions are not clearly codified.

In Hollywood, it was rare that the film composer ever met with or spoke to anyone on the postproduction sound crew. Both New York and the Bay Area had more instances of the film composer collaborating with the sound designer with the end soundscape reflecting the sensibilities of both.

Three Distinctive Soundscapes

It is worth noting that the three films, *Barton Fink*, *Bram Stoker's Dracula*, and *The English Patient*, have narratives that are reflective of the very film cultures from where they hail. *Barton Fink*, the most obvious, is about a New York playwright, and has a gritty New York style. It is understated in its visual design and has only the sound elements necessary to drive the narrative. The musical score is minimal, and plays supporting player to the sound design. There are no extra "frills" to *Barton Fink* as a film. One should not take from this statement that all New York films are "lean and mean," for certainly Martin Scorsese's *The Age of Innocence* (1993) is grand in scope, but even for Scorsese, Lievsay's sound design is quiet and reflective and does not have the "hyper-opulence"[4] that it might had it been "posted"[5] in Hollywood. Instead, it reflects the constrained morals and manners depicted in the story.

Bram Stoker's Dracula has all the familiar trappings of a major Hollywood film. It is lush in its visual splendor, the soundtrack is omniscient, and makes use of every possible device. There is little in the narrative that does not have some kind of sonic referent. This is most definitely, of the three films, the most lavish and studio-like, even though it was a Coppola film, which always implies unpredictability and independence.

The English Patient is the most inaccessible as a narrative. The film is long, the plot is complicated, and it has the "English actors and English director" cachet. It is high art,[6] which is what might be expected from the San Francisco Bay Area, as a bit of a cliché. The elite character of the film—indeed about elites—lends itself to sound minimalism. This particular soundscape is the most understated, delicate, and pristine, and is also high art—like a fine wine or delicate sculpture.

Admittedly, these three soundtracks deserve more attention than can be offered here, but they provide a glimpse of how disparate sound designs can be astounding in differing ways. The final sound mixes of *Barton Fink*, *Bram Stoker's Dracula*, and *The English Patient* are arguably three of the more artfully rendered. The clarity and specificity demanded of every sound selected, performed, and edited resulted in mixes that are emotionally connecting and provide their own sonic narratives. It comes as no surprise that the three award-winning films with their notable soundtracks would be creative and artful. What might be surprising is how much these films and their respective soundtracks reflect the very cultures of the professionals interviewed. Each of the aural styles of the soundscapes are distinctive. From the independent and humorful style of the Coen brothers in New York, to the industrial "shot on the lot" and inspired empowerment by Coppola for Sony, to the elegant and restrained Minghella with Saul Zaentz, the three soundtracks, and the films themselves, resemble the work style and even the communication style of the professionals—artists—I interviewed.

The Interviews

While it is illuminating to discover the various ways that the three film cultures are revealed in the work practices and the approaches to sound design, what is additionally intriguing is the contrast in the very interviews with the individuals. The New York sound professionals interviewed were more interested in discussing the end result than recalling the process. It was not easy to obtain amplification regarding the specific day-to-day experiences, nor were any of the principals able to recall how they approached the work, or many of the specific issues regarding how the technological challenges arose. Past publications that were printed closer to the release of the film were more useful for reflections regarding the specific "process" experiences. What the New York professionals seemed to connect with was the relationship with the film, the Coens—who they all enjoy tremendously—and the freedom of having an independent facility. The result is what concerned these professionals. Lievsay, more than anyone, was interested in the process of creativity and feeling free to innovate without limitations from others. One could argue that this is what most artists might focus on.

The Hollywood professionals preferred to talk about process. The workflow is of key importance to this group because they think in terms of time—what needs to get done by what day in the schedule. This group seems to be essentially temporally and process driven. The soundtrack they create has to fit within a certain amount of time. Unlike other creatives in the Hollywood system, who are indulged with more time if they need it, Hollywood postproduction sound professionals are not as likely to be perceived as needing time to be creative. Stone and Cohn both spoke in terms of schedules and making things happen within a specific period of time. However, Stone was remarkable in his ability to remember everything each individual contributed. He keeps detailed notebooks about each film he supervises and has spotting notes from the director, and copies of cue sheets for every aspect of the soundtrack. His preference toward artistic collaboration was evident when I interviewed him.[7]

Pat Jackson and Mark Berger were the most aesthetically involved in my interviews with them. These Bay Area professionals were very intellectual and articulate. They had theories, methods, and conceptual ideas for every aspect of the soundtrack. Even over twenty years after the film has been released, they were artistically vivid in their descriptions. Jackson was more detail oriented and articulate regarding specific process issues, while Berger was abstract and conceptual. They are both closely aligned with Murch, and their interest in the artistry becomes abundantly clear when reading their reflections about *The English Patient*, or sound design in general.

While no sweeping conclusions should be drawn from three films, with these three groups of professionals, it is safe to glean some contrasts between the fast-paced New York flows, the industrial Hollywood work practices, and the San Francisco theoretical analyses, and these contrasts are reflected in the discussions. The New York professionals work in an urban setting and there is a lack of nonsense or frill in the discussions about process or methods. The Hollywood professionals are practical and remember details about technological challenges that are specific and humorous. The Bay Area professionals are more conceptual and the interviews are long and theoretical. What must not be overlooked is the synthesis of the three cultures over the subsequent decades that has occurred as they share unions, hardware, and productions in common presently.

Collective Sound Design

The assertion that only a few key sound designers warrant attention as particularly creative or especially innovative should be revisited. This is not what is revealed through this investigation with these film sound professionals, nor do I believe these three case studies support that idea. With each of the three films what is revealed, rather, is three examples of collective authorship where one person has been assigned the task to lead a team of contributing creators.[8] Additionally, in each of these three films, the directors are reliant on the sound design as the aural imagination of the narrative. Even the ultimate sound designer himself, the person responsible for the title—Walter Murch—relied on not only another gifted artist to envision and develop a soundtrack—Pat Jackson—but she also relied on others to contribute their creativity and innovation to co-construct a mesmerizing soundtrack.

The notion of collective authorship—which Caldwell addresses from the perspective of the "industrial auteur"—necessarily refers to a product, which develops with the contributions of many, yet the illusion of a single driver of the process perseveres.[9] The single driver, in this case, is the supervising sound editor, but in reality, it is the entire sound team.[10] David Stone elaborates, "I really dislike the myth of the sound designer as a one-man band. It just doesn't work that way. The best thing a supervising sound editor can do is facilitate the contribution of all of the crew members with these disparate pieces into a coherent whole which is then finessed by the mixers."[11]

The flaw in the postproduction sound process, as it is typically practiced, is its money-conscious sensibility that insists each professional work in tandem with each other, without orchestration and with little time to reflect and refine. This workflow practice is one key reason why sound professionals are seen as technical workers, rather than the creative artists they actually are. Stone argues it well:

> What would be ideal in sound editing (and we have not achieved this) is if everyone involved could hear what everyone else has been doing so this exchange can feed and influence each other's work. What we have always done is work independently so no one hears all the parts come together until the final mix is in progress.[12]

What Stone describes is a cultural difference between the more industrial nature of postproduction sound than what occurred in the Bay Area and New York. What was valued in Hollywood was the most efficient and economical method of soundtrack delivery. However, those who chose to work in the Bay Area did so specifically to avoid that lifestyle. The ability to take more time and collaborate more fluidly was a key motivation for the Bay Area workers. Additionally, Saul Zaentz would ensure they got the time necessary for any projects he produced.[13] As for sound professionals in New York, while efficiency is valued and schedules are not luxurious, directors are loyal to their sound supervisors, who in turn are loyal to their crews. "There is an efficiency that comes with anticipating how each other is going to work."[14]

Final Thoughts

In each of the case studies, what becomes clear is the sound professionals adapted the digital editing tools to their own purposes. Whether their reasons were for creative freedom, for integrating analog and digital platforms, or for advancing expediency and efficiency, the agency and innovation of each individual and each group of workers dominated. The technology utilized was socially constructed and adapted for the needs and desires of the professional sound editors to allow them maximum creativity and minimum technological obstruction.

While addressing the cultural differences between three film sound communities during the initial transitions into digital sound editing in feature filmmaking, what is important to remember is that eventually, all three communities eventually adopted Pro Tools as the industry standard. Ironically, the digital platform slowest to enter the market with any serious competitive edge became the format most adaptable to the needs of the postproduction sound workers. As the professionals required changes and new applications, Pro Tools proved to be the more responsive format, with the best technical support.[15] Rather than illustrate the unstoppable direction of technology, what is revealed is the power of personal agency when adapting tools to the task at hand. Indeed, the three case studies provide examples of three distinct instances of creative sound designers, who adapted their technologies for their own uses, with varied purposes and results.

Only in Hollywood do artificial designations like "above-the-line" or "below-the-line"[16] even enter a conversation with any regularity. Future scholarship would do well to look past these constructed categories and see the specific contributions of postproduction sound professionals as creatives who utilize and adapt their technologies rather than rely on the assumption that the technologies define the craft. Additionally, it seems clear that relying on a few celebrity sound mavens as indicative of the overall trajectory of sound design in its entirety ignores the various distinctions that clarify the art of sound design from the craft of executing the editing and mixing. The art form is too complex, too collective, and too labor intensive to ever really know where the ideas and ingenuity actually come from. I would submit that Coppola said it best when discussing his process for *Bram Stoker's Dracula* as one of "gathering the mushrooms for this pie."[17]

Notes

1 Here, referring to Taylorism as it developed in the late nineteenth century used to shorten training times and skills requirements for jobs.
2 I worked at both Skywalker and Zaentz for two years, 1987–9, and worked alone that entire time as a Foley artist.
3 In 1995, Larry Blake, supervising sound editor for *Underneath* (1995), insisted Foley artist Alicia Stevenson be allowed to work with Marko Costanzo in New York. This is according to Bruce Pross, who states that this was the only time he recalls this happening and that it was unusual and awkward. Bruce Pross, Personal Interview, interview by Vanessa Theme Ament, Skype, in New York, June 10, 2014.
4 Hollywood would normally be considered the "hyperreal" culture. New York has a reputation of a more naturalistic aesthetic, even if stylized.
5 This is industrial jargon, which is shorthand for having had its postproduction completed.
6 This term "high art" is always debatable, but can be applied to *The English Patient* as a more traditionally "British Style" epic film.
7 As I have known Stone since 1982, and had been married to him, but also worked on over twenty films and television show with him, his process is well known to me.
8 In the appendix are the credit lists of the sound professionals who worked on *Barton Fink*, *Bram Stoker's Dracula*, and *The English Patient*, as exhibits of the collective authorship in sound design.

9   John Thornton Caldwell, *Production Culture: Industrial Reflexivity and Critical Practice in Film and Television* (Durham, NC: Duke University Press, 2008), 199.
10 After receiving their Academy Award for Best Sound Effects, Tom McCarthy, Jr. and David Stone split their time with the reflexivity required of successful sound professionals: McCarthy thanked the upper-level executives at Sony, and Stone named every contributing sound professional personally (personal observation, industrial ethnography).
11 David E. Stone, Personal Interview, by Vanessa Theme Ament, Skype, in Savannah, GA, November 20, 2010.
12 Ibid.
13 Malcolm Fife, Personal Interview, by Vanessa Theme Ament, Phone, in San Francisco, CA, June 1, 2014.
14 Bruce Pross, Personal Interview.
15 The three initial formats, Post Pro, Wave Frame, and Sonic Solutions, have not been dominantly used in sound design since Pro Tools became the industry standard in the late 1990s.
16 Recall the discussion on "above" and "below-the-line" in Chapter 1.
17 This is repeated from Chapter 4. Francis Ford Coppola and Kim Aubry, *Bram Stoker's Dracula* (Sony Pictures Home Entertainment, 2007).

Appendix A

Workflow Diagram for *Barton Fink*

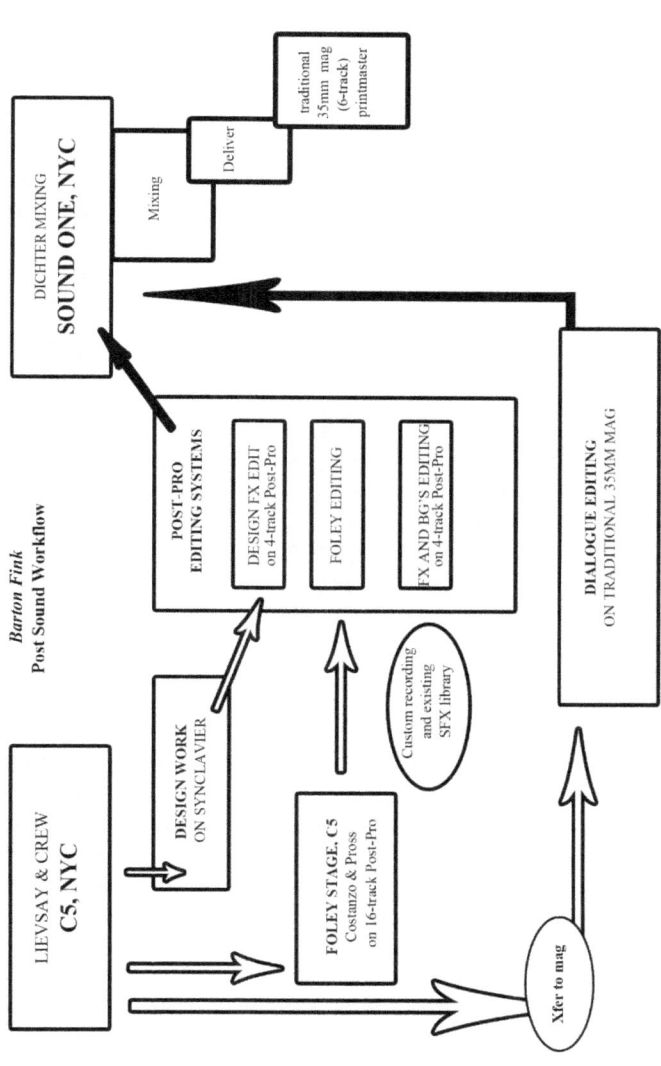

Appendix B

Workflow Diagram for *Bram Stoker's Dracula*

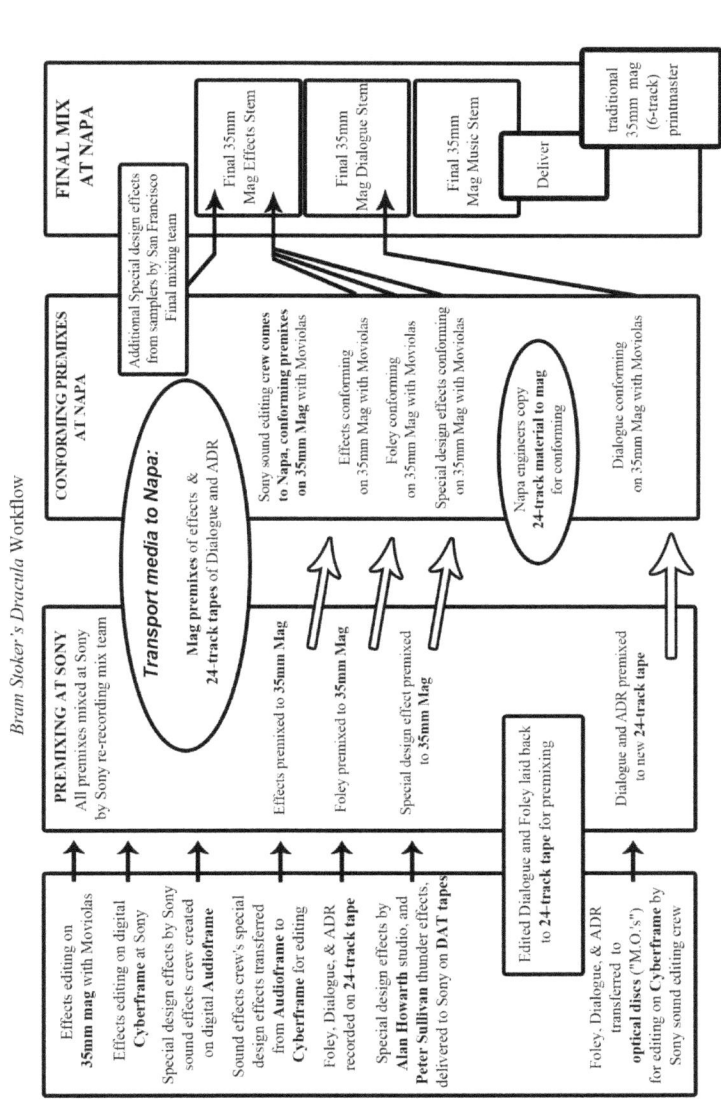

Appendix C

Workflow Diagram for *The English Patient*

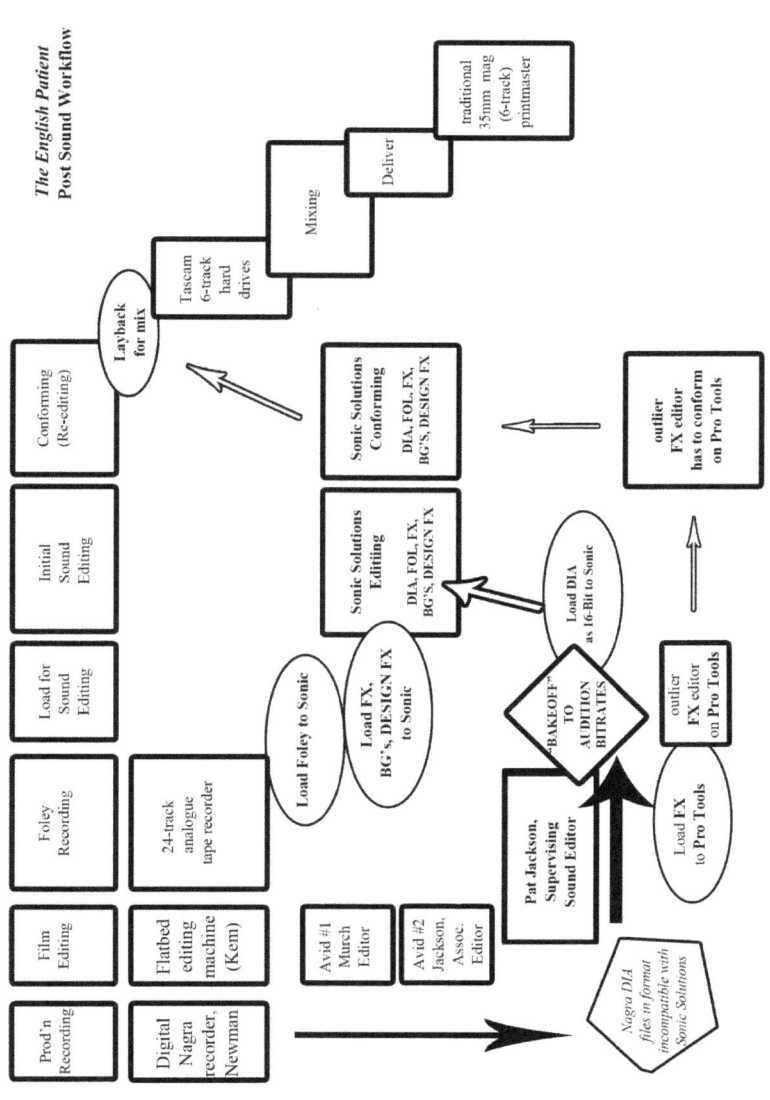

Appendix D

Sound Department Credits

In Alphabetical Order (from IMDb.com)

Barton Fink (1991)

Allan Byer	Production sound mixer
Jean Marie Carroll	Sound recordist
Missy Cohen	Apprentice sound editor
Randall Coleman	Assistant sound editor (as Randy Coleman)
Marko A. Costanzo	Foley artist
Lee Dichter	Rerecording mixer
William Docker	Assistant sound editor (as Bill Docker)
Harry Higgins	Sound rerecordist
Bradford L. Hohle	Stereo sound consultant: Dolby
Kerry Kelley	Sound rerecordist
Frank Kern	Sound editor
Peter F. Kurland	Boom operator
Blake Leyh	Sound editor
Skip Lievsay	Supervising sound editor
Marissa Littlefield	Sound editor
Bruce Pross	Foley supervisor
Anne Sawyer	Assistant sound editor
Gail Showalter	ADR editor
Philip Stockton	Dialogue supervisor
Dominick Tavella	Foley mixer
Steven Visscher	Sound editor

Bram Stoker's Dracula (1992)

John H. Arrufat	Dialogue editor (as J. H. Arrufat)
Christopher S. Aud	Sound effects editor
Steve Borne	Assistant sound editor (as Steve Born)
Zane D. Bruce	Foley walker
Harry Cheney	Dialogue editor (as Harry M. Cheney)
Kim B. Christensen	Assistant sound designer: American Zoetrope
David B. Cohn	Supervising ADR editor
Roy Finch	Temp sound mix editor: American Zoetrope
Linda Folk	ADR editor
Jeff Gomillion	ADR mixer
Mark Gordon	Supervising foley editor
Alan Howarth	Additional sound effects
Robert Janiger	Production mixer
Devin Joseph	Assistant sound editor
Mark R. LaPointe	Dialogue editor (as Mark LaPointe)
Catt LeBaigue	Foley editor (as Catherine H. LeBaigue)
Mark L. Mangino	Sound effects editor
Cindy Marty	Foley editor
Tom C. McCarthy	Supervising sound editor (as Tom C. McCarthy MPSE)
Jim McKee	Special sound effects editor: American Zoetrope
Frances K. Petrovic	Assistant sound editor
Sanford Ponder	Sound effects editor
John F. Reynolds	Foley editor (as John Reynolds)
Aaron Rochin	Rerecording mixer
Diana J. Rogers	Apprentice sound editor
Joseph T. Sabella	Foley walker (as Joe Sabella)
Dennis S. Sands	Rerecording mixer (as Dennis Sands)
Jackson Schwartz	Foley mixer
George W. Scott	Boom operator
Leslie Shatz	Rerecording mixer/sound designer
John Sisti	ADR editor (as John L. Sisti)
Greg Stewart	Sound effects editor (as Gregory C. Stewart)

David E. Stone	Supervising sound editor (as David E. Stone MPSE)
David F. Van Slyke	Sound effects editor (as David Van Slyke)
Marian Wallace	Rerecording mixer
David Williams	Sound effects editor
David W. Alstadter	ADR mixer (uncredited)
Kim Aubry	ADR mixer (uncredited)
Gary Boggess	Assistant sound effects editor (uncredited)
Ed Callahan	Sound editor (uncredited)
Tina Canny	Machine room operator (uncredited)
Derek Casari	Dubbing engineer (uncredited)
Jeremy J. Gordon	Sound editor (uncredited)
Larry Hopkins	Layback sound mixer (uncredited)
Andy Murdock	Sound recordist (uncredited)
Andy Newell	Sound effects editor (uncredited)
Timothy Pearson	Foley artist (uncredited)
Philip Rogers	Sound recordist: final mix fixes (uncredited)
B. Tennyson Sebastian III	Rerecording mixer (uncredited)

The English Patient (1996)

James Austin	Staff: The Saul Zaentz Film Center (as Jim Austin)
Michael Axinn	Apprentice sound editor
Don Banks	Boom operator (as Donald Banks)
David Franklin Bergad	Assistant dialogue editor
Mark Berger	Rerecording mixer
Sara Bolder	Dialogue editor
Loren Byer	Staff: The Saul Zaentz Film Center
Frank Canonica	Staff: The Saul Zaentz Film Center
Vince Casper	Staff: The Saul Zaentz Film Center
Amanda Chan	Staff: The Saul Zaentz Film Center
Tobin Davis	Assistant dialogue editor (as Tobin Delaca Davis)
Adriano Di Lorenzo	Sound assistant
Adam Dornbusch	Sound department intern
Richard Duarte	Foley mixer

Malcolm Fife	Foley editor
Grant Foerster	Staff: The Saul Zaentz Film Center
Steve Fontano	Assistant foley mixer
Anna Geyer	Staff: The Saul Zaentz Film Center
Aura Gilge	Assistant sound effects editor (as Aura Belle Gilge)
Pat Jackson	Supervising sound editor
Rick Kahn	Staff: The Saul Zaentz Film Center
Stephen Kearney	Assistant sound effects editor
Michael Kelly	Staff: The Saul Zaentz Film Center
Mark Levinson	ADR supervisor
Scott Levitin	Staff: The Saul Zaentz Film Center
Kyrsten Mate	Sound effects editor (as Kyrsten Mate Comoglio)
Kathy McVey	Staff: The Saul Zaentz Film Center
Marnie Moore	Foley artist
Walter Murch	Rerecording mixer
Douglas Murray	Sound effects editor (as Douglas S. Murray)
Jennifer Myers	Foley artist
Christopher Newman	Sound recordist (as Chris Newman)
John Nutt	Dialogue editor
Margie O'Malley	Foley artist
Dan Olmsted	Staff: The Saul Zaentz Film Center
David Parker	Rerecording mixer
Jim Pasque	Staff: The Saul Zaentz Film Center
Mark Paul	Staff: The Saul Zaentz Film Center
Richard Quinn	ADR editor
Scott G. Roberts	Staff: The Saul Zaentz Film Center (as Scott Roberts)
David Roesch	Staff: The Saul Zaentz Film Center
Gianni Sardo	Sound mixer: second unit
Roy Segal	Staff: The Saul Zaentz Film Center
Ivan Sharrock	Sound recordist
Steve Shurtz	Staff: The Saul Zaentz Film Center
Greg Simon	Staff: The Saul Zaentz Film Center
Dianna Stirpe	Dialogue editor
Marc-Jon Sullivan	Boom operator (as Marc Jon Sullivan)
Diego Taborda	Sound department intern

John Timperley	Sound engineer
Joe Tysl	Staff: The Saul Zaentz Film Center
Jennifer L. Ware	Sound effects editor
Laurie Wentworth	Staff: The Saul Zaentz Film Center
Jeff Whittle	Staff: The Saul Zaentz Film Center
Mary Works	Assistant dialogue editor
Marilyn S. Zalkan	First assistant sound editor
Liberata Zocchi	Sound transfer: Rome
Frank Canonica	Supervising sound recordist: mix stage (uncredited)
Ann Kroeber	Additional sound effects (uncredited)
Luciano Muratori	Sound supervisor (uncredited)
Primiano Muratori	Sound supervisor (uncredited)
Douglas Murray	Sound designer (uncredited)
Michael Semanick	Rerecording mixer: premixes (uncredited)
Peter Steinbach	Additional sound effects (uncredited)
Steven Yeaw	Transfer operator (uncredited)

Bibliography

Books and Articles

Allen, Robert, and Douglas Gomery. *Film History: Theory and Practice*. 1st ed. New York: McGraw-Hill Humanities/Social Sciences/Languages, 1993.

Altman, Rick. *Silent Film Sound*. New York: Columbia University Press, 2004.

Altman, Rick. *Sound Theory, Sound Practice*. New York: Routledge, 1992.

Altman, Rick. *A Theory of Narrative*. New York: Columbia University Press, 2008.

Altman, Special Editor Rick. *Yale French Studies 60: Cinema/Sound*. New Haven, CT: Yale French Studies, 1980.

Ament, Vanessa, T. *The Foley Grail: The Art of Performing Sound for Film, Games, and Animation*. Amsterdam; Boston, MA: Focal Press/Elsevier, 2009.

Ament, Vanessa, T. *The Foley Grail: The Art of Performing Sound for Film, Games, and Animation*. 2nd ed. Burlington, MA: Taylor & Francis, 2014.

Ament, Vanessa, T. "Leo Chaloukian Interview, Cont." *MovieSound Newsletter* 2, no. 1 (February 1992): 2–3.

Ament, Vanessa, T. "Sound Director Runs TV Academy." *MovieSound Newsletter* 1, no. 19 (November 1991): 1–2.

Bachelard, Gaston. *The Poetics of Space*. 1st ed. Boston, MA: Beacon Press, 1994.

Balio, Tino, ed. *The American Film Industry*. Rev. ed. Madison: University of Wisconsin Press, 1985.

Barnes, Randall. "The Sound of Coen Comedy: Music, Dialogue and Sound Effects in *Raising Arizona*." *Soundtrack* 1, no. 1 (March 2008): 15–28. doi:10.1386/st.1.1.15/1.

Baron, Cynthia, Diane Carson, and Frank P. Tomasulo, eds. *More Than a Method: Trends and Traditions in Contemporary Film Performance*. Detroit: Wayne State University Press, 2004.

Barthes, Roland. *Image-Music-Text*. Translated by Stephen Heath. New York: Hill and Wang, 1978.

Barthes, Roland. *The Pleasure of the Text*. Translated by Richard Miller. Reissue. New York: Hill and Wang, 1975.

Bazin, Andre. *What Is Cinema? Vol. 1*. Translated by Hugh Gray. Rev. Ed. Berkeley: University of California Press, 2004.

Beck, Jay, and Tony Grajeda, eds. *Lowering the Boom: Critical Studies in Film Sound*. Champaign: University of Illinois Press, 2008.

Benjamin, Oscar. "The English Patient." *Film Score Monthly* 2 (1997): 25.

Bernard, H. Russell. *Research Methods in Anthropology: Qualitative and Quantitative Approaches*. 4th ed. Lanham, MD: AltaMira Press, 2005.

Bertini, Louis. "The Busy Boom Years of Local 771." *Editors Guild Magazine* 1, no. 5 (October 2012): 1.

Bijker, Wiebe E., Thomas P. Hughes, and Trevor J. Pinch, eds. *The Social Construction of Technological Systems: New Directions in the Sociology and History of Technology*. Cambridge, MA: MIT Press, 2012.

Biskind, Peter. *Easy Riders Raging Bulls: How the Sex-Drugs-And Rock 'N Roll Generation Saved Hollywood*. 1st Touchstone ed. New York: Simon & Schuster, 2011.

Bordwell, David, Janet Staiger, and Kristin Thompson. *The Classical Hollywood Cinema: Film Style & Mode of Production to 1960*. New York: Columbia University Press, 1985.

Bourdieu, Pierre. *Distinction: A Social Critique of the Judgement of Taste*. Translated by Richard Nice. Cambridge, MA: Harvard University Press, 1984.

Bourdieu, Pierre. *The Field of Cultural Production*. Edited by Randal Johnson. New York: Columbia University Press, 1993.

Buckland, Warren, ed. *Film Theory and Contemporary Hollywood Movies*. 1st ed. New York: Routledge, 2009.

Buhler, James, David Neumeyer, and Rob Deemer. *Hearing the Movies: Music and Sound in Film History*. New York: Oxford University Press, 2009.

Byrge, Duane, and Mike Barnes. "Legendary Producer Saul Zaentz Dies at 92." *Hollywood Reporter*, January 2, 2014, 1.

Caldwell, John. *Production Culture: Industrial Reflexivity and Critical Practice in Film and Television*. Durham, NC: Duke University Press Books, 2008.

Campbell, Joseph. *The Power of Myth*. New York: Anchor, 1991.

Carey, James W. *Communication as Culture, Revised Edition: Essays on Media and Society*. New York: Routledge, 1992.

Caughie, John. *Theories of Authorship*. New ed. New York: Routledge, 1981.

Champlin, Charles. *George Lucas: The Creative Impulse*. Rev. upd. Su ed. New York: Harry N. Abrams, 1997.

Chang, Heewon. *Autoethnography as Method*. Walnut Creek, CA: Left Coast Press, 2009.

Chion, Michel. *Audio-Vision: Sound on Screen*. Translated by Claudia Gorbman. New York: Columbia University Press, 1994.

Chion, Michel. *Film, a Sound Art*. Translated by Claudia Gorbman and C. Jon Delogu. New York: Columbia University Press, 2009.

Chion, Michel. "The Man Who Was Indeed There (Carter Burwell and the Coen Brothers' Films)." *Soundtrack* 1, no. 3 (November 2008): 175–81. doi:10.1386/st.1.3.175/1.

Chion, Michel, Claudia Gorbman, and Walter Murch. *Audio-Vision: Sound on Screen.* New York: Columbia University Press, 1994.

Cooke, Mervyn. *A History of Film Music.* Cambridge: Cambridge University Press, 2008.

Coppola, Francis Ford. *The Conversation.* Lions Gate, 1974.

Coppola, Francis Ford, and James V. Hart. *The Making of Bram Stoker's "Dracula."* 1st ed. London: Pan Books, 1992.

Coppola, Francis Ford, and Kim Aubry. *Bram Stoker's Dracula.* Sony Pictures Home Entertainment, 2007.

Costin, Midge. *Making Waves: The Art of Cinematic Sound.* Ain't Heard Nothin' Yet Corp. and Goodmovies Entertainment, 2019.

Crafton, Donald. *The Talkies: American Cinema's Transition to Sound, 1926–1931.* 1st ed. Oakland: University of California Press, 1999.

Deuze, Mark. *Managing Media Work.* 1st ed. Thousand Oaks, CA: Sage, 2010.

During, Simon, ed. *The Cultural Studies Reader.* 3rd ed. New York: Routledge, 2007.

Franklin, Peter, "*King Kong* and Film on Music: Out of the Fog," in *Film Music: Critical Approaches*, ed. Kevin J. Donnelly. Edinburgh: Edinburgh University Press, 2001, 88–102.

Friedman, Ted. *Electric Dreams: Computers in American Culture.* New York: New York University Press, 2005.

Gay, Paul du, Stuart Hall, Linda Janes, Hugh Mackay, and Keith Negus. *Doing Cultural Studies: The Story of the Sony Walkman.* 1st ed. London: Sage, 1997.

Gilman, Charlotte Perkins. *The Yellow Wallpaper and Other Stories.* Unabridged ed. Mineola, NY: Dover, 1997.

Gitlin, Todd. *Inside Prime Time: With a New Introduction.* 1st ed. Oakland: University of California Press, 2000.

Gomery, Douglas. *The Coming of Sound.* London: Routledge, 2005.

Gorbman, Claudia. *Unheard Melodies: Narrative Film Music.* 1st printing. Bloomington: Indiana University Press, 1987.

Greiving, Timothy. "All's Well That's Burwell: Part 1." *Film Score Monthly (FSM Online)* 13, no. 9 (2008): 11.

Griffin, Nancy, and Kim Masters. *Hit and Run: How Jon Peters and Peter Guber Took Sony for a Ride in Hollywood.* New York: Simon & Schuster, 1997.

Gunning, Tom. "Cinema of Attractions," in *Encyclopedia of Early Cinema*, ed. Richard Abel. London: Routledge, 2005, 124–7.

Hainge, Greg. "The Unbearable Blandness of Being: The Everyday and Muzak in Barton Fink and Fargo." *Post Script* 27, no. 2 (winter 2008): 38–47.

Hall, Stuart, Dorothy Hobson, Andrew Lowe, and Paul Willis, eds. *Culture, Media, Language.* London: Hutchinson, 1980.

Henkin, David. *City Reading: Written Words and Public Spaces in Antebellum New York.* New York: Columbia University Press, 1998.

Hesmondhalgh, David. *The Cultural Industries.* London: Sage, 2012.

Hesmondhalgh, David, and Sarah Baker. *Creative Labour: Media Work in Three Cultural Industries.* 1st ed. New York: Routledge, 2011.

Hill, John, and Pamela Church Gibson, eds. *The Oxford Guide to Film Studies.* New York: Oxford University Press, 1998.

Hitchcock, Alfred. *The Birds*, 2013.

Holman, Tomlinson. *Sound for Digital Video.* Boston, MA: Focal Press, 2005.

Holman, Tomlinson. *Sound for Film and Television.* 3rd ed. Amsterdam; Boston, MA: Focal Press/Elsevier, 2010.

Holt, Jennifer, and Alisa Perren. *Media Industries: History, Theory, and Method.* 1st ed. Chichester: Wiley-Blackwell, 2009.

Horkheimer, Max, and Theodor W. Adorno. *Dialectic of Enlightenment.* Edited by Gunzelin Schmid Noerr. Translated by Edmund Jephcott. 1st ed. Palo Alto, CA: Stanford University Press, 2007.

Ihde, Don. *Listening and Voice: Phenomenologies of Sound.* 2nd ed. Albany: State University of New York Press, 2007.

Jung, C. G. *The Archetypes and the Collective Unconscious.* Translated by R. F. C. Hull. 2nd ed. Princeton, NJ: Princeton University Press, 1981.

Katz, Susan Bullington. "A Conversation with ... Anthony Minghella." *Written by: The Journal of the Writers Guild of America* 1 (March 1997): 22–8.

Keil, Charlie. *Early American Cinema in Transition: Story, Style, and Filmmaking, 1907–1913.* 1st ed. Madison: University of Wisconsin Press, 2002.

Kenny, Tom. *Sound for Picture.* 2nd ed. Vallejo, CA: Artistpro, 2000.

Kerins, Mark. *Beyond Dolby (Stereo): Cinema in the Digital Sound Age.* Bloomington: Indiana University Press, 2010.

Kipen, David. *The Schreiber Theory: A Radical Rewrite of American Film History.* Brooklyn, NY: Melville House, 2006.

Kirschenbaum, Jill. "The Coen Connection." *Millimeter* 19 (May 1991): 72.

Knapp, Bettina L. *A Jungian Approach to Literature.* 1st ed. Carbondale: Southern Illinois University Press, 1984.

Kolker, Robert. *A Cinema of Loneliness.* 4th ed. New York: Oxford University Press, 2011.

Laing, Heather. *Gabriel Yared's The English Patient: A Film Score Guide.* Lanham, MD: Scarecrow Press, 2007.

Leitch, Thomas. *Film Adaptation and Its Discontents: From "Gone with the Wind" to "The Passion of the Christ."* Baltimore, MD: Johns Hopkins University Press, 2009.

Lewis, Jon. *Whom God Wishes to Destroy . . .: Francis Coppola and the New Hollywood.* Durham, NC: Duke University Press, 1997.

Lewis, Jon, and Eric Smoodin, eds. *Looking Past the Screen: Case Studies in American Film History and Method*. Durham, NC: Duke University Press Books, 2007.

Lippy, Tod, ed. *Projections 11: New York Film-Makers on New York Film-Making*. London: Faber & Faber, 2000.

LoBrutto, Vincent. *Sound-On-Film: Interviews with Creators of Film Sound*. Westport, CT: Praeger, 1994.

Madison, D. Soyini. *Critical Ethnography: Method, Ethics, and Performance*. 1st ed. Thousand Oaks, CA: Sage, 2005.

Mayer, Vicki. *Below the Line: Producers and Production Studies in the New Television Economy*. Durham, NC: Duke University Press Books, 2011.

Mayer, Vicki, and Caldwell. *Production Studies: Cultural Studies of Media Industries*. New York: Routledge, 2009.

Miller, Toby. *The Well-Tempered Self: Citizenship, Culture, and the Postmodern Subject*. Baltimore, MD: Johns Hopkins University Press, 1993.

Minghella, Anthony, Ralph Fiennes, Juliette Binoche, Kristen Scott Tomas, Michael Ondaatje, and Buena Vista Home Video (Firm), Miramax Home Entertainment (Firm). *The English Patient*. Burbank, CA: Miramax Home Entertainment: Distributed by Buena Vista Home Video, 1998.

Mitchell, Donald. *Cultural Geography: A Critical Introduction*. 1st ed. Malden, MA: Blackwell, 2000.

Morris, Earl. *The Fog of War: Eleven Lessons from the Life of Robert S. McNamara*. Sony Classics, 2003.

Muncey, Tessa. *Creating Autoethnographies*. Los Angeles: Sage, 2010.

Murch, Muriel. "Conversation with Walter Murch and Michael Ondaatje." *Projections*, 1998, 311–26.

Murch, Walter. "How Do You Like Your Room? Thoughts on the Use of Sound in *Barton Fink*." *Soundtrack* 1, no. 3 (November 2008): 211–15. doi:10.1386/st.1.3.211/1.

Murch, Walter. *In the Blink of an Eye*. Rev. 2nd ed. City: Silman-James Press, 2001.

Nancy, Jean-Luc, and Charlotte Mandell. *Listening*. Annotated ed. New York: Fordham University Press, 2007.

Ondaatje, Michael. *The Conversations: Walter Murch and the Art of Editing Film*. New York: Knopf, 2004.

Ondaatje, Michael. *The English Patient*. New York: Vintage Books, 1993.

Padilha, José. *Robocop*. MGM (Video & DVD), 2014.

Pallasmaa, Juhani. *The Eyes of the Skin: Architecture and the Senses*. 2nd ed. Waltham, MA: Academy Press, 2005.

Powdermaker, Hortense. *Hollywood: The Dream Factory*. 1st ed. New York: Little, Brown, 1950.

Prochnik, George. *In Pursuit of Silence: Listening for Meaning in a World of Noise*. 1st ed. New York: Doubleday, 2010.

Rose, Jay. *Audio Postproduction for Film and Video, Second Edition: After-the-Shoot Solutions, Professional Techniques, and Cookbook Recipes to Make Your Project Sound Better*. 2nd ed. Boston, MA: Focal Press, 2008.

Rosten, Leo. *Hollywood: The Movie Colony, the Movie Makers*. New York: Harcourt Brace, 1942.

Rubin, Michael. *Droidmaker: George Lucas and the Digital Revolution*. 1st ed. Gainesville, FL: Triad, 2012.

Ruhlmann, William. "Skip Lievsay." *Premiere* 5 (October 1991): 41–2.

Rushing, Janice Hocker, and Thomas S. Frentz. *Projecting the Shadow: The Cyborg Hero in American Film*. 1st ed. Chicago: University of Chicago Press, 1995.

Schatz, Thomas. *Hollywood Genres: Formulas, Filmmaking, and the Studio System*. 1st ed. New York: McGraw-Hill Humanities/Social Sciences/Languages, 1981.

Schatz, Thomas, and Steven Bach. *The Genius of the System: Hollywood Filmmaking in the Studio Era*. Minneapolis: University of Minnesota Press, 2010.

Schechner, Richard. *Performance Theory*. 2nd rev. ed. New York: Routledge, 2003.

Schelle, Michael. *The Score: Interviews with Film Composers*. Los Angeles: Silman-James Press, 1998.

Scott, Allen J. *On Hollywood: The Place, the Industry*. Princeton, NJ: Princeton University Press, 2005.

Shank, Barry. *Dissonant Identities the Rock "N" Roll Scene in Austin, Texas*. Hanover, NH: University Press of New England, 1994. http://site.ebrary.com/id/10477847.

Smillie, Grahame. *Analogue and Digital Communication Techniques*. 1st ed. London: Butterworth-Heinemann, 1999.

Staiger, Janet. *Interpreting Films*. Princeton, NJ: Princeton University Press, 1992.

Sterne, Jonathan. *The Audible Past: Cultural Origins of Sound Reproduction*. Durham, NC: Duke University Press Books, 2003.

Stoker, Bram. *Dracula*. New York: Dover, 2000.

Stoker, Bram. *The New Annotated Dracula*. Edited by Leslie S. Klinger. Annotated. New York: W. W. Norton, 2008.

Stone, David. "1993 Is Digital Summer." *MovieSound Newsletter* 3, no. 1 (Summer): 1–2.

Taubin, Amy. "A Sound Is Built." *Village Voice* 34 (October 3, 1989): 71–2.

Thornton, Sarah. *Club Cultures: Music, Media, and Subcultural Capital*. Hanover: Wesleyan, 1996.

Viers, Ric. *The Location Sound Bible: How to Record Professional Dialog for Film and TV*. Michael Wiese Productions, 2012.

Weis, Elisabeth, and John Belton, eds. *Film Sound: Theory and Practice*. New York: Columbia University Press, 1985.

Whittington, William. *Sound Design and Science Fiction*. Austin: University of Texas Press, 2007.

Wierzbicki, James, Nathan Platte, and Colin Roust, eds. *The Routledge Film Music Sourcebook*. London: Routledge, 2011.

Winston, Brian. *Media Technology and Society: A History from the Telegraph to the Internet*. Reissue. New York: Routledge, 1998.

Wyatt, Justin. *High Concept: Movies and Marketing in Hollywood*. Austin: University of Texas Press, 1994.

Yewdall, David Lewis. *Practical Art of Motion Picture Sound*. Waltham, MA: Focal Press, 2012.

Online Sources

"1978 New England Digital Synclavier." *Mix Magazine*, September 1, 2006, 1.

"1987 Sonic Solutions NoNoise." *Mix Magazine*, September 1, 2006, 1.

"A Brief History of Pro Tools." Accessed May 30, 2014. http://www.musicradar.com/tuition/tech/a-brief-history-of-pro-tools-452963/.

"Academy Awards Database." *Oscars.org*. Accessed May 30, 2014. awardsdatabase.oscars.org/ampas_award/DisplayMain.jsp;E2(C45743F519FECC4E92C5A2474E2FD?curTime=1401413545144.

AP. "Richard Vorisek, 71, Film Sound Director." *New York Times*, November 9, 1989, sec. Obituaries. http://www.nytimes.com/1989/11/09/obituaries/richard-vorisek-71-film-sound-director.html.

Black, Dave. "Percussive Arts Society." *PAS Hall of Fame: Murray Spivack*, n.d. http://www.pas.org/experience/halloffame/SpivachMurray.aspx.

Deitrick, William R. "Automated Computer Controlled Editing Sound System (ACCESS)." In *Proceedings of the May 19–22, 1980, National Computer Conference*, 83–5. AFIPS '80. New York: ACM, 1980. doi:10.1145/1500518.1500532.

"Digital F/X Announces Acquisition of Waveframe; Merges into New Digital F/X Audio Digital F/X." The Free Library. *PR Newswire*, September 30, 1992.

EditDroid NAB 2. Video. Las Vegas, NV. Accessed May 30, 2014. http://www.youtube.com/watch?v=27hVFq2RB1w.

"Editors Guild Magazine Article: The Busy Boom Years of Local 771." *MPEG*. Accessed June 15, 2014. https://www.editorsguild.com/magazine.cfm?ArticleID=1133.

"Evan Brooks." *NAMM.org*. Accessed May 30, 2014. http://www.namm.org/library/oral-history/evan-brooks.

Ferguson, Kevin L. "Bright Lights Film Journal." *Bright Lights Film Journal*. Accessed June 17, 2014. http://brightlightsfilm.com/.

"Guild's History." *MPEG*. Accessed June 15, 2014. https://www.editorsguild.com/Guildshistory.cfm.

Inc, InfoWorld Media Group. *InfoWorld*. InfoWorld Media Group, Inc., 1979.

"Mark Berger Faculty Page U.C. Berkeley." University. *Department of Film and Media U.C. Berkeley*, June 4, 2014. http://fm.berkeley.edu/people/faculty/149-2/.

Moorer, James. "James A. Moorer Resumé," May 30, 2014. http://www.jamminpower.com/main/resume.html.

"Roy Segal Retires: Head of Fantasy Studios, Saul Zaentz Film Center," May 1, 2000, 1.

Silverman, Leon. "The New Post Production Workflow: Today and Tomorrow." *Motion Kodak*. Accessed May 30, 2014. http://www.motion.kodak.com/motion/uploadedFiles/US_plugins_acrobat_en_motion_hub_Post_Production2.pdfhttp://www.motion.kodak.com/motion/uploadedFiles/US_plugins_acrobat_en_motion_hub_Post_Production2.pdf.

"Sonic Solutions NoNOISE Honored with Emmy Award; Recognized for Groundbreaking Digital Sound Restoration Product." *Business Wire*, October 2, 1996.

Staff. "Sound One Corp." *Mix Magazine*, September 1, 2000. http://www.mixonline.com/mag/audio_sound_one_corp/.

Xu, Jodi. "Sonic Acquires DivX for about $326 Million." *Wall Street Journal*, June 2, 2010. http://online.wsj.com/article/SB10001424052748703561604575282812506317640.html.

Personal Interviews

Berger, Mark. Interview by Vanessa Theme. Skype in Berkeley, CA, July 17, 2013.

Birnbaum, Elisha. Interview by Vanessa Theme Ament. Phone in New York, September 22, 2013.

Cohn, David B. Interview by Vanessa Theme Ament. Skype in Simi Valley, CA, October 23, 2013.

Costanzo, Marko. Interview by Vanessa Theme Ament. Skype in Fort Lee, NJ, June 23, 2013.

Dichter, Lee. Interview by Vanessa Theme Ament. Phone in New York, June 17, 2014.

Fife, Malcolm. Interview by Vanessa Theme Ament. Phone in San Francisco, CA, June 29, 2013.

Fife, Malcolm. Interview by Vanessa Theme Ament. Phone in San Francisco, CA, June 1, 2014.

Fleischman, Thomas. Interview by Vanessa Theme Ament. Skype in New York, June 23, 2013.

Fulmis, Jim. Edited by Vanessa Theme Ament. E-mail, June 3, 2014.
Fulmis, Jim. Interview by Vanessa Theme Ament. In person, Studio City, CA, March 10, 2008.
Gearty, Eugene. Interview by Vanessa Theme Ament. Skype in Beaufort, SC, May 22, 2014.
Jackson, Pat. Interview by Vanessa Theme Ament. Skype in San Francisco, CA, July 18, 2013.
Kroeber, Ann. Interview by Vanessa Theme Ament, July 24, 2013.
Lee, Steve. Private message, November 9, 2019.
Lievsay, Skip. Interview by Vanessa Theme Ament. Phone in Los Angeles, CA, August 18, 2013.
Lievsay, Skip. Interview by Vanessa Theme Ament. Phone in New York, June 10, 2014.
McCarthy Jr., Thomas. Interview by Vanessa Theme Ament. In person, November 15, 2012.
Pross, Bruce. Interview by Vanessa Theme Ament. Skype in New York, June 10, 2014.
Shurtz, Steve. Interview by Vanessa Theme Ament. Facebook private message, May 29, 2014.
Stone, David E. Interview by Vanessa Theme Ament. In person, in Chicago, IL, February 9, 2008.
Stone, David E. Interview by Vanessa Theme Ament. Skype in Savannah, GA, November 20, 2010.
Stone, David E. "Sound Design Scene Descriptions from Dracula," November 20, 2010.

Index

Note: Page numbers with "n" indicates endnotes in the text.

20th Century Fox Studios 100
24-track machines 26, 43n.5, 67n.26, 83, 106

above-the-line 6, 59, 129
Academy Award for Best Sound 30, 32, 100
Academy Award for Best Sound Editing 28
Academy Award for Best Sound Effects Editing 72, 93n.31, 130n.10
Academy of Motion Picture Arts and Sciences 26, 41, 47n.62, 117n.29
acousmatic 62, 81–82, 95n.67
adaptation 8, 72–75, 118n.42
Age of Innocence, The (Scorsese) 124
Aladdin (film) 76, 93n.31
Allen, Woody 11, 33
alternative to analog 50–52
Altman, Rick 6
Amadeus (film) 30
American Graffiti (Coppola) 3
American Zoetrope (production company) 9–10, 28, 85, 103
analog editing 53; *see also* editing
analog postproduction sound editing 8
Anderson, Richard L. 76
Andrews, Naveen 101
Apocalypse Now (Coppola) 2–3, 10, 21n.44, 33, 75, 93n.36
atypical collaboration 54–55
Aud, Christopher S. 80
Audioframe 78–80
aural narrative 6, 15, 17, 111
Automated Computer Controlled Editing Sound System (ACCESS) 41–43
automated dialogue replacement (ADR) editors 26, 32, 40, 78–79, 82–85, 102
AVID (software) 38, 102, 107
Avid Technologies 39

Bachelard, Gaston 64
Baker, Chet 31
Barton Fink (Coen) 36, 43, 49–66, 124
  atypical collaboration 54–55
  creating an independent workspace 52–53
  embracing an alternative to analog 50–52
  getting noticed 59–61
  musical chairs 58–59
  sound evokes the uncanny 56–58
  stepping up to the Foley plate 55–56
  workflow diagram 131
Batman Returns (film) 76
Beggs, Richard 77, 94n.38
below-the-line film professional 5–8, 59, 129
Bender, William 35
Berger, Mark 19n.8, 30, 32, 45n.33, 106, 108–112
big sound films 10, 29, 42; *see also* film(s)
Binoche, Juliette 100
Birds, The (film) 3, 19n.10
Birnbaum, Elisha 34–35
Biskind, Peter 9, 18n.2
Blood Simple (Coen) 36
Blue Velvet (film) 30
Bochar, Ron 35, 51, 56
Boekelheide, Jay 31
Bram Stoker's Dracula (BSD) (film) 28, 38, 71–91, 124
  Academy Award for Best Sound Editing 28
  adaptation 72–75
  collaboration 80–82
  Columbia 75–76
  Coppola, Francis Ford 75–76
  reediting 82–85

revisiting 82–85
revoicing 82–85
Shadow 85–91
Sony 75–76
technology 78–80
workflow diagram 133
Brill Building 34
Brooks, Evan 40
Burtt, Ben 4, 30, 48n.74
Burwell, Carter 36, 49, 54, 58–59, 65

c5 Sound 43, 52, 55–56, 60, 66–67n.9, 67n.29
Caldwell, John T. 5–6, 127
California 28, 42, 77, 102, 109
Campbell, Chuck 43n.4
Canton, Mark 76
Caul, Harry 2
celebrity sound designers 4–5, 12
Celluloid Closet, The (film) 104
Chaloukian, Leo 27
Champlin, Charles 9
Chazelle, Damien 4
Chelsea Five 66n.9
Chew, Richard 103
Chion, Michel 82, 95n.67, 97n.93
Christensen, Kim B. 80, 84
Coen, Ethan 11, 34, 122
Coen, Joel 11–12, 34, 122
Coen brothers 4, 20n.20, 36, 49–66, 125; *see also Barton Fink* (Coen)
Cohen, Emory 37
Cohn, Dave B. 79, 82–85, 126
"Collaborating Editor Clause" 35
collaboration 80–82
  of artists 59
  atypical 54–55
  authorship and 23n.67
  cooperative 65
  cross-pollination and 93n.36
  between music and sound design 115
  sound and music teams 84
  technology 16, 36
  "working without a net" 92n.9
collective sound design 126–128; *see also* sound design
"Color Eats Sound" 109–110

Columbia Pictures 11, 71–72, 75–77, 85
Concord Music Group 30
"consumption junction" 14, 22n.65
Convergence Corporation 39
Conversation, The (film) 1–2, 60–61, 103
Coppola, Francis Ford 1–3, 9–10, 18n.4, 19n.7–8, 21n.44, 28, 32–33, 71–77, 83–84, 92n.13–14, 93n.23, 96n.80, 103, 125
Costanzo, Marko 50, 55–58, 61, 65, 66n.3, 129n.3
Cotton Club, The (film) 75
Cowan, Schwartz 14
creative artists 6–7, 17, 121, 127
Creedence Clearwater Revival 31
Crutcher, Ethel 59, 68n.53
Cyberframe/WaveFrame (PC-based editorial system) 37–38, 78–80, 83, 94n.47, 104
Cybermation 37

DAT 79–81
de Almásy, László 101–102, 110–113
Defoe, Willem 101
Deitrick, William R. 41
Demme, Jonathan 11, 33
Dichter, Lee 53, 58
Digidesign 40
Digidrums 40
digital audio workstation (DAW) 36–38, 42–43, 60
digital editing 27, 50, 52, 101
  nascent incarnation 27
  Sonic Solutions 39
  sound 11–14, 36, 41, 60, 66, 121–122, 128
  system 37–38, 41, 50, 66
  tools 128
  transition 13, 17, 43, 55, 60, 94n.40, 101, 104, 122
  venture 11
Digital F/X 38
digital revolution 38
Digital Workstation Post Pro 35
digital workstations 21n.37, 41, 94n.47
Direct to Disk ("D to D") model 42–43
disconnect—sound professionals 6

Dolby Laboratories 26, 83
Dolgen, Jonathan 76, 78
Dolgenism 79
Doris, Robert J. 39
Dracula (film) 28, 93n.31
Droidmaker: George Lucas and the Digital Revolution (Rubin) 9, 18n.4
DroidWorks 39
Drumulator (digital drum machine) 40

Easy Riders Raging Bulls: How the Sex-Drugs-And Rock 'N Roll Generation Saved Hollywood (Biskind) 9, 18n.2
EditDroid 38–41, 48n.74
editing 43n.3
  analog 53
  electronic 16, 27, 38, 44n.10, 53, 60, 94n.47
  reediting 82–85
  technology 66n.3
Egypt 101
Electric Dreams : Computers in American Culture (Friedman) 14
Electric Sound and Picture (ESP) system 38
electronic sound editing *see* sound editing
Elisabeta 74
Emmy for Technical Achievement 39
England 115
English Patient, The (film) 30, 32, 41, 99–116, 124
  Academy Award for Best Sound 32, 100
  "Color Eats Sound" 109–110
  conceptualizing 101–103
  ephemeral dualities 110–116
  executing 101–103
  Foley Follies 106–107
  generalist as specialist 103–104
  Sonic Solutions 104–106
  workflow diagram 135
ephemeral dualities 110–116

Fairlight 38, 106
Fantasy Records 9, 30–32, 40, 45n.29, 104
Ferguson, Kevin L. 63
Fiennes, Ralph 101

Fife, Malcolm 106
film(s)
  aural narratives of 6
  big sound 10, 29, 42
  following television 37–38
  magnetic 25, 38, 40, 43, 43n.3, 43n.5, 57, 78–79, 81, 84, 94n.48, 96n.78–79, 104, 106, 116n.10
  score 3–4, 7, 49, 58, 89, 113
  sound 2, 4, 6–9, 11–13, 18, 25–27, 30, 32, 34, 40, 42–43, 46n.37, 52, 60, 69n.64, 100, 121
  sound professionals 3, 6
filmmaking 2, 5, 9–10, 13, 21n.38
  art-centered 33
  conventions 5
  corporate conglomerates 92n.18
  culture 2, 9
  documentary 103
  industrialized nature of 10
  intellectualism 118n.52
  low-budget 36
  system 21n.38
  technology 26
  tech-savvy neophyte editors 72
  theatrical and literary traditions 34
Fincher, David 4
Firth, Collin 101
flatbed 101, 116n.10
Fleischman, Tom 34
Flick, Steve 4
Foley
  artists 6, 16, 20n.20, 25–26, 29, 43n.4, 44n.18, 51, 67n.29, 78, 107, 123, 129n.3
  Follies 106–107
  mixer 46n.42, 47n.69, 50, 55
  mixing 38, 40, 106
  stages 26, 29, 35, 38, 43n.4–5, 51, 56–58, 66n.3, 67n.29
Folk, Linda 84
Fordist assembly line method 20n.22
Fordist Hollywood System 25–28
"for hire" 5, 20n.21
Franklin, Peter 7, 20n.32
freelance sound professionals 10
Friedman, Edward 14

Fuchs, Fred 77
Fulmis, Jim 37

Gabriel, 106, 108
Galas, Diamanda 75, 82
Gearty, Eugene 12, 41–43, 50–51, 56, 60
Geisler, Ben 58
generalist as specialist 103–104
George Lucas: The Creative Impulse (Champlin) 9
Glassman, Remi 19n.10
Godfather, The (Coppola) 28, 73, 75
Goodfellas (Lievsay) 59
Goodman, John 57
Gotcher, Peter 40
Grateful Dead, The (musician) 32
"Great Man" theory 19n.18
Grindstaff, Chuck 37
Grindstaff, Doug 37
Guber, Peter 75

"hang it as a unit" 25, 43n.1
Harker, Jonathan 74, 81–82, 86–88
Hart, James V. 72–73
Hastings Sound 11, 35
Helsing, Van 74, 89–90
Herrmann, Bernard 3
Hesmondhalgh, David 7, 21n.33
Heuer, Ellen 43n.4
Hollywood 21n.8, 22n.55, 61, 75–78, 122–127
  above-the-line and below-the-line 59, 129
  A-List 77
  Blacklist 34
  community 2, 18n.4
  conversion from analog to digital 9, 11, 50, 72
  divisions 37
  electronic technology 12
  filmmaking conventions 5
  filmmaking process 13
  industrial conventions 6
  industry professionals 38
  Local 776 33
  New York sound community 60
  postproduction sound crews 51
  sound technologies 100
  studio tradition 10
Hollywood, the Dream Factory (Powdermaker) 13
Hollywood: The Movie Colony, the Movie Makers (Rosten) 13
Holt, Skip 37
Hotel Earle 56, 61–63, 65
Howarth, Alan 80–81, 86

Iatrou, Mildred 4, 19n.17
IBM 37
independent workspace 52–53
Indiana Jones and the Last Crusade (film) 77
industrial ethnography, transition through 12–15
International Alliance for Theatrical Stagehand Employees (IATSE) 27, 33, 35
Italy 100

Jackson, Pat 31, 99–100, 102–106, 108–111, 116, 117n.28, 117n.30, 118n.52, 118n.58, 126–127
Joplin, Janis 32

Kilar, Wojciech 75, 84
King, Richard 4
King Kong (film) 7, 20n.32
Klyce, Ren 4
Kohut, Mike 83

Laser Pacific (postproduction sound facility) 37
Lee, Ai-Ling 4, 19n.17
Lerner, Michael 58
Libya 101
Lievsay, Skip 4, 12, 35–36, 42–43, 49–54, 57–61, 64–66, 68n.54, 122, 124
LoBrutto, Vincent 13
Lucas, George 1–2, 4, 9–10, 18n.4, 27–30, 33, 38–39
Lumet, Sidney 33
Lynch, David 31

magnetic film 25, 38, 40, 43, 43n.3, 43n.5, 57, 78–79, 81, 84, 94n.48, 96n.78–79, 104, 106, 116n.10; *see also* film(s)

Mangini, Mark 4, 76, 93n.30
mapping digital transition 8–12
Marin County 9, 28–30, 39
McCarthy, Tom 76–77, 79, 93n.32, 130n.10
McDormand, Frances 34
MGM 75, 79, 81
Miller's Crossing (Coen) 50–51, 53
Minghella, Anthony 100–101, 107–108, 110, 123, 125
Mini-Micro Systems, Inc. 41
Miramax (independent distribution firm) 34, 101
"mixing in the box" 60, 68n.57
Moore, Demi 100
Moorer, James A. 39
Moorer, Jim 104
Morita, Yoshiko 76
Motion Picture Sound Editors (MPSE) 59, 68n.53
Moving Pictures Technicians Local 16 33
Mulligan, Gerry 31
Multitrack Sonic System 39
Mulvehill, Charles 77
Murch, Walter 1–4, 18n.1, 18n.4, 19n.6, 28, 31–32, 77, 94n.38, 99–103, 105–107, 110–111, 127
Murray, Douglas 31
musical chairs 58–59
musical score 74, 97n.81, 124
myth of postproduction sound 5–8

National Association of Music Merchants (NAMM) 40
National Endowment of the Arts 34
National Organization of Broadcasters 38–39
Neiman-Tillar Associates 41
Neroda, Emil 35, 42–43
New England Digital Corporation 35, 42–43, 50
New Hollywood 2
New Line (independent distribution firm) 34
Newman, Chris 105–106
New York 5, 9, 11, 21n.39, 22n.55, 33–36, 51, 59–60, 72, 99–100, 102, 122–123, 125–126

sound design 12
Nichols, Mike 11, 33
Nolan, Christopher 4
nondiegetic film score 49; *see also* film(s)
NoNoise˚ (Macintosh-based system) 39

O'Connell, Kevin 6
Ondaatje, Michael 100
One Flew Over the Cuckoo's Nest (Zaentz) 10
One from the Heart (film) 75
"On Wallpaper in Some Films" (Ferguson) 63
Outsiders, The (film) 75

Parker, David 106, 110
Pearl, Phil 35
Penn, Arthur 33
Peters, Jon 75
Peters, Val 35
Ponder, Sanford 80–82
Pospisil, John 3
Post Audio Production (PAP) 37
Post Pro 55, 60
Post Pro (hard drive editing system) 50, 56–57
postproduction sound 5–8, 10–12, 121
  companies 11
  editing 36
  professionals 6–8, 14, 71, 125–126, 129
Powdermaker, Hortense 13
Pross, Bruce 35, 55–58, 65, 66n.3, 129n.3
prosumer equipment 3, 19n.15
Pro Tools 32, 39–40, 100, 104–107, 122

Quittner, Katherine 75, 84, 97n.81

Raging Bull (film) 11
Rain People, The (Coppola) 2
random-access memory (RAM) 42
reassessing sound design *see* sound design
reediting 82–85, 123
Reeves, Keanu 76, 82, 96n.70, 97n.89
Repulsion (film) 61
repurposing 84, 97n.81
revisiting 82–85
revoicing 82–85
Right Stuff, The (film) 30

Robocop (film) 3, 26
Roesch, John 43n.4
Rosten, Leo 13
rough edit 102, 116n.14
Rubin, Michael 9, 18n.4
Rumble Fish (film) 75
Ryder, Winona 73, 76, 81, 93n.22
Rydstrom, Gary 4, 30

Sable, Dan 35
Sala, Oskar 19n.10
San Francisco Bay Area 1, 5, 9–12, 21n.40, 28, 40–41, 51, 60, 66, 76, 99–100, 102, 104, 122–123, 127–128
   expansion 32–33
Sauer, Mary C. 39
Saul Zaentz Film Center 9–10, 29–32, 77, 99, 104, 106, 122
Sayles, John 33
Schamus, James 34
Scientific and Engineering Award 41
Scorsese, Martin 11, 33, 124
Secret Garden, The (film) 104
Segal, Roy 32, 39–41, 104–107, 122
Shadow 85–91, 97n.92
Shatz, Leslie 31, 77–78, 80, 82, 84, 86
Shurtz, Steve 40
Silence of the Lambs, The (Lievsay) 59
Sisti, John 79
Skywalker Ranch (movie ranch) 9, 29–30, 39, 44n.18, 94n.38, 104
Skywalker Sound 77
Social Construction of Technology (SCOT) theory 14
Sonic Solutions (digital editing platform) 39–41, 104–106
Sony 11, 28, 71–72, 75–80, 83, 93n.25, 125
sound
   consultant 3
   designers 2–5, 12, 61, 126
   editors 3, 10–11, 15, 26–27, 29, 32–33, 44n.17, 51, 109, 123
   effects 3–4, 7, 26, 31, 37, 40, 51, 54, 58, 81, 84
   engineer 7
   mixers 33
   mixing 121
   myth of postproduction 5–8
   postproduction 5–8, 10–11
   professionals 6–8, 10–11, 14–15, 19n.16, 21n.37, 125, 128
sound design 1–5, 28, 49, 51, 56, 71, 78, 85, 91, 115–116, 121–129
   collective 126–128
   cultural transition 121–123
   essence of 8
   interviews 125–126
   New York 12
   soundscapes 124–125
   technology 14
SoundDroid (digital sound editing system) 39
sound editing 8, 11, 105, 121
   analog postproduction 8
   digital 11–14, 36, 41, 60, 66, 121–122, 128
   transition from analog to digital 11
Sound One (sound facility) 11, 35–36, 42, 50–53, 56, 59
Sound-On-Film (LoBrutto) 13
soundscapes 124–125
Sound Shop (postproduction sound facility) 35, 42–43
Sound Tools (digital system) 40–41
soundtrack 4–7, 11–12, 16–17
   big and loud 68n.54
   delivery 128
   digital 52
   films 71
   film score 113
   Foley and ADR 23n.67
   Foley artists' ability 25
   imaginative 42
   New York 59
   omniscient 124
   postproduction 30
   Skywalker 30
   vitality and exuberance 27
Soundtrack 54
Spear, Guy 35
Special Achievement Award 26
Spivack, Murray 7
Splet, Alen 31–32, 42, 46n.37
Sprocket Systems 39

Steiner, Max 7
Steven, Richard 37
Stockton, Phil 35, 51
Stone, David E. 73, 76, 78–82, 84–85, 93n.30, 96n.80, 97n.82, 122, 126–127, 130n.10
symbol creator 7
symbol maker 7
synchresis 7, 21n.36
Synclavier digital synthesizer 42–43, 50–52, 56–57
synthesist 3, 95n.54
synthesizers 3, 16, 19n.13, 38, 42, 80

technological geography 36–37
technology
  adaptation 14
  Bram Stoker's Dracula (BSD) (film) 78–80
  collaboration 16, 36
  digital 11, 52, 99, 102
  editing 66n.3
  editorial 100
  electronic 12
  of film sound 6
  machines and 8
  sound design 14
  transition 72
  WaveFrame 38
television sound editors/editorial 11, 26–27
Thom, Randy 4, 30
Thomas, Kristin Scott 100
THX 26, 29

Todd-AO (post-production company) 38
Toys (film) 77
Trans Audio Sound 11, 35
Trans Audio Video (TAV) 41–42
transition
  from analog to digital sound editing 11
  to digital editing 101, 122
  into digital sound editing 121
  mapping digital 8–12
  technology 72
  through industrial ethnography 12–15

Unbearable Lightness of Being, The (film) 30
uncanny 56–58, 119n.60
United States 100

Van Slyke, David F. 80
Vlad the Impaler (Dracula) 73–74
Vorisek, Dick 34–35, 53, 123
Vorisek, Jack 35

Wally Heider Studios 32
Warner, Frank 11, 22n.55
Warner Brothers 38
Warner Hollywood 93n.29
Who Framed Roger Rabbit? (film) 43n.4
Willow (film) 77

Yared, Gabriel 107
Young Indiana Jones (television series) 39

Zaentz, Saul 10, 100, 123, 125, 128
Zoetrope (film studio) 77

www.ingramcontent.com/pod-product-compliance
Lightning Source LLC
Chambersburg PA
CBHW070640300426
44111CB00013B/2192